Positive Psychology in the Military

Studies in Military Psychology and Pedagogy

Edited by Hubert Annen

Vol. 15

Allister MacIntyre, Danielle Charbonneau and
Hubert Annen (eds.)

Positive Psychology in the Military

Lausanne • Berlin • Bruxelles • Chennai • New York • Oxford

Bibliographic Information published by the Deutsche Nationalbibliothek
The Deutsche Nationalbibliothek lists this publication in the Deutsche Nationalbibliografie; detailed bibliographic data is available online at http://dnb.d-nb.de.

Library of Congress Cataloging-in-Publication Data
A CIP catalog record for this book has been applied for at the Library of Congress.

ISSN 1619-778X
ISBN 978-3-631-89413-2 (Print)
E-ISBN 978-3-631-89414-9 (E-PDF)
E-ISBN 978-3-631-89415-6 (EPUB)
DOI 10.3726/b20427

Open Access: This work is licensed under a Creative Commons Attribution CC-BY 4.0 license. To view a copy of this license, visit https://creativecommons.org/licenses/by/4.0/

© Allister MacIntyre, Danielle Charbonneau and Hubert Annen, 2023

Peter Lang Group AG
International Academic Publishers
Bern

This publication has been peer reviewed.

info@peterlang.com - www.pterlang.com

Contents

Allister MacIntyre
Foreword 7

Samir Rawat and Abhijit Deshpande
Enhancing Functional Fitness through Self – Regulation Initiatives in the
Military: An Indian Perspective 9

Madlaina Niederhauser and Hubert Annen
Resilience and Resilience Training: Focus on Military Science 27

Liisi Toom
Built-in Resilience Intervention in Academic Military Education 55

Walter L Giusti and Gayle Sherwell
From Military to Government Sector Applications: Developing an
Organizational Climate Framework to Promote Psychological Health,
Well-being and Positive Work Behaviors 75

Petrus C. Bester
A Positive Psychology Perspective on Predeployment Fitness-For-Duty
Evaluations for External Deployments: A Proposition for the South
African National Defence Force 107

Nity Sharma and Vrishti Kapoor
Flourish: Promoting Positivism and Psychological Well-Being
in Military 127

Valerie Wood and Lobna Chérif
Appreciating the Strengths of Comrades: A Positive Psychology
Intervention for Improved Military Leadership, Unit Cohesion, and
Member Well-Being 143

Mala Agarwal
Mindfulness Tools to Combat Stress in Military Personnel 155

Tyler E. Freeman
Stress Mindset as an Enabler of Soldier Well-being 173

Edith Knight
Meaningful Work: What Does It Mean? 195

Martin I. Jones, Sophie L. Wardle, and Fiona N. Koivula
And So to Bed: Sleep, Well-being, and Human Performance 215

Jürgen Leon and Elena Trentini
When Others Have Your Back: Relational Leadership in Balint Style
Groups to Enhance Stress Coping Strategies among Armed Forces 237

Verma Swati, Updesh Kumar, and Dakshi Walia
Post Traumatic Growth in Military Personnel 263

The Authors 277

Allister MacIntyre

Foreword

I am very excited and honored to offer a foreword for this volume on positive psychology and mental health. These are undeniably uncertain times for people throughout the world and the uplifting theme of this publication is timely. Although the COVID-19 situation has been improving, we have not fully escaped the grip of this devastating pandemic. The virus continues to mutate with new variations of the disease cropping up on a consistent basis. Furthermore, our collective attention is riveted by disturbing global events and worldwide unrest including political turmoil, humanitarian crises, human rights violations, significant protest activities, natural disasters, climate change, economic woe and recession, terrorist activities, immigration and refugee issues, civil wars, military conflicts, and even full-scale war. This list is by no means exhaustive. Additionally, thanks to the proliferation of access to the internet, the real time dissemination of these crises reaches public awareness at a level unheard of before. Unfortunately, the sources of information on the internet are not always reliable. It is widely known that social media also propagates falsehoods, twisted facts, unsubstantiated conspiracy theories, and outright fear mongering. This profusion of misinformation contributes to widespread heightened stress levels and anxiety. It is safe to assume that, within the military populations of the world, we can expect highly elevated levels of personal tension.

Given that VUCA (volatility, uncertainty, complexity, and ambiguity) is a reality on the world stage, it is clear that military forces must regularly grapple with situations where VUCA plays a prominent role, and deals with potentially harmful consequences. It is also a certainty that these serving members must be able to cope effectively with the inevitable stressors associated with these volatile circumstances or suffer significant physical and psychological consequences. It is impressive that 23 international scholars from eight different countries have taken the time to provide well-reasoned and empirically sound guidance on how to best navigate these ubiquitous and pervasive uncertainties while safeguarding their mental and physical health and well-being. Although there are no universal solutions, or silver bullets, the advice presented in the contents of this volume offers hope and wisdom to allow our serving members to emerge victorious despite unreasonable pressures and challenges.

Subjects covered in this publication include the importance of service fitness (Rawat & Deshpande) where the authors not only provide a comprehensive *Functional Fitness* model, but they also explain how self-awareness, exploration and regulation can contribute to this fitness. Given the theme of this volume, it is no surprise that attention is devoted to related topics. Hence, themes like resilience (Niederhauser & Annen; Toom), psychological health (Giusti & Sherwell), and positive psychology (Bester; Sharma & Kapoor; Wood & Cherif) play a prominent role in the guidance offered. Naturally, the dangers faced by military members are not purely physical and the cognitive role played by our minds is given the attention it deserves (Agarwal; Freeman). Edith Knight focuses on the satisfaction that one can derive from meaningful work, and Martin Jones, along with his colleagues Sophie Wardle and Fiona Koivula, examine restful sleep "as a pillar of well-being and human performance." No volume on military resilience and well-being would be complete without touching on the role played by leaders. With this in mind, Jürgen Leon and Elena Trentini explore how a relationship leadership approach can help to enhance stress coping strategies in military members. Finally, Verma Swati, Updesh Kumar and Dakshi Walia provide an optimistic view of what might emerge from stressful consequences. They readily acknowledge that it does not matter how much training soldiers receive, they will inevitably experience traumatic situations. However, the consequences of these negative events, is not always devastating. The authors explain that Post Traumatic Growth can occur when the trauma generates positive outcomes. This may sound counter-intuitive but they do explain what this can look like and describe how this growth can take place.

In summary, the world is an uncertain and complex place, filled with volatile and ambiguous situations. Moreover, military members must face these traumatic hardships more frequently than the citizens within their respective nations. However, the message need not be one of doom and gloom. The scholars contributing to this volume have emerged from our global village to provide sage advice on how to meet these adversities with resilience and perhaps even growth. It is clear that various cultures, and geographical boundaries, do not determine the quality of a human's response to suffering. Nor does it influence well-being, psychological health, resilience and growth. A single volume will never be able to capture every conceivable consequence of trauma, or provide perfect solutions, but the efforts offered by these gifted scholars provide a sense of hope.

Samir Rawat and Abhijit Deshpande

Enhancing Functional Fitness through Self – Regulation Initiatives in the Military: An Indian Perspective

Introduction

Functionally fit[1] leaders are likely to be more effective than leaders struggling with personal issues that may interfere with professional competence, critical decision making and genuine care and concern for soldiers under their command. They are those leaders who are not just physically fit, but mentally robust as well with sound psychological health to endure highly stressful and demanding military situations. To always ensure the functional fitness of troops for operational readiness and mission effectiveness, leaders need to be functionally fit and be able to take care of themselves first through the 'thick and thin of combat'.[2] This implies that leaders possess good cognitive abilities and personal awareness along with high self-regulation skills for regular introspection and reflection on one's own areas of improvement.[3]

Executive functions[4] refer to aspects of cognition that are called for in situations when brain and behavior cannot rely on automatic responses. They describe interrelated cognitive abilities that are activated when one must intentionally or deliberately process and retain information, manage, and integrate this information, and resolve competing or discrepant stimulus representations and response options. In other words, these cognitive abilities are important for military leaders to help them organize information, plan and problem solve. Furthermore, executive functions orchestrate the thoughts and actions necessary for goal-directed behavior that allows for purposeful engagement with the environment, even under conditions of duress. As such, they are aspects of the psychological ability that assists individuals in self-regulation and self-control.

This chapter focuses on self-regulation initiatives including optimal self-care to ensure functional fitness especially in operations that require military leaders to show courage. First, we define concepts related to functional fitness in the military as well as discuss the idea of self-regulation and its importance to military leaders. Second, we discuss the importance of holistic health, which includes physical and mental components like self-care. We argue that self-care provides structure and meaning to one's personal and professional life, both of which interact, impact, and complement each other. Third, we discuss various dimensions of self–regulation including self–awareness, self–exploration, relaxation, meditation, recreation and self enhancement through such self-exploratory activities as regularly diary writing as well as reading to upgrade one's knowledge, skills or fine-tune one's attitudes. Finally, the chapter defines and outlines the importance of intrapersonal self–awareness, interpersonal self–awareness, cultural self–awareness, communication self–awareness, and professional self–awareness.

Functional Fitness in the Military

The Functional Fitness model[5] outlines the multidimensional nature of service fitness. The model is comprehensive, integrating psychological and physical fitness dimensions into four psycho-physiological clusters: (a) physically unimpaired with depleted psychological resources; (b) physically impaired with depleted psychological resources; (c) physically impaired without depletion in psychological resources; and (d) physically unimpaired with actual performance boosted due to applied psychological resources. The model suggests that when functional decline occurs, the type of resource depletion should be identified before deciding the most appropriate interventions.

Interestingly, Parmak[6] proposed a multidimensional approach model to military health, which argues that functional fitness and health go beyond medical symptoms, and that it is necessary to examine performance restrictions from both a physical and a psychological perspective. In a discussion during WARMCAMP[7] 2016 and 2017 in India, the first author and Parmak noted that, even for cadets, the masculine and fitness-focused male-oriented military culture may unintentionally trigger a functional decline when cadets choose to inadvertently suppress, ignore or minimize some health issues (for example, a torn ligament) in order to be perceived as strong, fit and capable of passing their mandatory tests. The non-medical Military Demand – Resource model recognizes the role of soldiers' psychological resources while considering situational demands (e.g., routine military tasks, daily stressors, risks), soldiers' internal and external resource environment (e.g., leadership, coping mechanism, operational tasks), and outputs (performance, training standards, promotion), all of which mediate situational demands and outcomes. This model has been studied by other res earchers[8,9,10,11,12] with similar findings.

Self-Regulation and Its Importance to Military Leaders

In this chapter, *self-regulation*[13] is defined as the process of purposefully directing one's actions, thoughts, and feelings towards a goal; it includes methods and means involved in achieving and sustaining goals, conceptualized as internally represented desired states. At the core of most self-regulation theories[14,15,16,17] is a feedback loop that starts with individuals setting goals, then comparing their progress against the goals, and finally modifying their behaviors or cognitions to reduce any discrepancy between the goals and the current state. Evidence suggests that leaders who display self-control[18,19,20] and monitor and adjust their behavior best respond to the complexity and demands of the environment.

Self-regulation theories[21,22,23] suggest that leaders who readily adapt to a variety of organizational requirements and contexts can maximize their effectiveness by putting 'service before self.' Self-regulation is what allows leaders to put the organizational interest first and delay their own gratification with respect to desired personal outcomes.[24,25] Self-regulation can promote trust[26] responsibility,[27] understanding of others,[28] leadership emergence[29] as well as cooperation with others.[30] Researchers have reported exciting findings about what makes leaders effective,[31] including (a) promoting self-awareness among group members, (b) setting clear standards and goals for their team, and (c) motivating team members to reduce the discrepancy between desired and actual performance. Similarly, other researchers[32] have identified three essential conditions for successful group regulation: (a) self-awareness, (b) clear standards and goals, and (c) the ability, willingness, and necessary resources to make changes.

Although important features like effort,[33] goal-directedness,[34] problem-solving[35] and persistence[36] have been associated with successful leadership, few studies have connected functional fitness with the self-regulation framework of effective leadership.

Significance and Value of Holistic Health in the Military

The military emphasizes soldiers' physical fitness for obvious reasons yet tends to disregard the increasingly recognized contribution of psychological processes to health and fitness, especially in the case of peak performance.[37] Evidence of the important role of behavior and emotion in holistic health is accumulating and military leaders should become knowledgeable on this topic. Health does not depend only on medical status but also depends on the environment and living conditions.[38]

Some researchers[39] view health as an individual attribute – a harmonious balance of physical and psychological well-being, which include an ample supply of physical resources, social activities and good interpersonal relations with others, a favorable disposition, attitude and temper, freedom of movement, lifestyle, and effectiveness in action.

Antonovsky[40] approached health from a coherence perspective and described it as the capacity to perceive one's world as significant and manageable. Some researchers[41] reported positive self-evaluations as predictors of good mental as well as physical health while others reported internal locus of control[42] and self-efficacy[43] as positive health indicators.

Health is often considered a self–responsibility in the military; optimal health can best be obtained by enhancing the health of individual soldiers; by strengthening the military community as a whole with improved access to facilities for enhancing physical and mental health as well as by encouraging a paradigm shift from physical fitness to holistic health. In sum, holistic health goes beyond just physical fitness and leaders need to recognize the critical role that attitudes, feelings, emotions and behavior may play in shaping a soldier health.

Perspectives in Self–Regulation

In the subsequent paragraphs, we will describe self-regulation using the constructs of self-awareness, self-exploration, relaxation, meditation, recreation, and inner work.

Self-Awareness

Self-awareness[44] skills include domains in personal and professional learning as well as development in the individual, interpersonal, cultural, and physical domains. It reflects the individual aspects that military leaders bring to their leadership style. For instance, leaders may focus their attention on the espoused military values, endorse them, and live by them. Leaders must be aware of their personal values to recognize when they may accommodate conflicting values and up to what point, when faced with making decisions under duress. Other relevant aspects of self-awareness include cultural sensitivity, commitment and competence which are impacted by one's congruence and which may likely reduce the discrepancy between the espoused military values and one's behavior, increase transparency and, in turn, acceptance as a true leader.

Senior leaders expect their staff officers to be supportive and to give them positive feedback. However, it requires courage for leaders to be receptive to less flattering feedback on both the personal and professional front. Although increasing self-awareness may temporarily raise self-consciousness and self-doubt, it will lead to positive outcomes if it is used for exploring one's own characteristics and clarifying one's own values. There are many possible ways to cultivate self –awareness and self-development, which military leaders may choose from. A few such strategies are mentioned in the following paragraphs, but the list is not exhaustive.

Self–Exploration

An important aspect of self-care skills includes self–exploration,[45] which is crucial to become more aware and less prone to overreactions. While emotional stability is achieved through the development of self-exploration, it may be beneficial professionally to help leaders thrive and flourish in becoming a better version of themselves. While self-awareness may represent the capacity of becoming the object of one's own attention by identifying, processing and storing information about self by reflecting on the experience of perceiving and processing stimuli (e.g., I am drinking carrot juice and this is healthy as well as tasty), self–exploration is the process of discovering that there is something innate and universal that helps us in resolving our inner conflicts by becoming aware of our natural acceptance (e.g., I have been consuming alcohol for many years now in what started as peer pressure to conform during college days. However, do I need to consider healthier lifestyles because I am not getting any younger? Plus, burning the calories from the momentary pleasures in being a 'good chap with tremendous capacity' to my weekend buddies is becoming a bigger challenge.).

In other words, self-awareness is more about knowing what and who you are while self– exploration gyrates towards reducing discrepancies between knowing who you are and what you can become.

Relaxation

The four most common strategies to achieve relaxation[46] are breathing exercises, mindfulness practice, relaxation exercises and guided imagery. While deep conscious breathing may induce a sense of deep relaxation and progressive muscle tension reduction, improper breathing is known to increase stress, tension and is associated with a feeling of muscle tightness.

Mindfulness[47] refers to the practice of stillness, centering attention on the breath and full awareness of the present moment; it requires concentration to bring one's awareness into focus and to direct attention to the present moment.

Relaxation exercises[48] are based on the principle that a body cannot be relaxed and anxious at the same time. As such, relaxation strategies aim at reducing anxiety, inducing sleep, relaxing the body to facilitate coping with specific anticipated events or reducing phobic reactions to certain events perceived as threatening.

Guided imagery[49] or visualization as it is also known is based on the principles of progressive relaxation and aids relaxation and concentration by involving all the senses.

Meditation

Meditation[50] is about finding a stillness within and being in the moment, while mindfulness refers to efforts made to bring one's awareness in the here and now. Leaders who meditate on a regular basis can more easily access their potential by tapping into their own resources while remaining aware of the external circumstances surrounding them. Lama Surya Das[51] provides an excellent summary of the benefits of a regular meditation practice:

- Brings greater calm, peacefulness, and awareness of inner resources
- Helps the mind remove clutter and confusion
- Slows down restless mind or mind filled with anger, hate, fearful thoughts and negative emotions
- Brings about centeredness and a sense of balance
- Makes senses and perceptions more powerful
- Assists in gaining insight in personal issues and greater understanding of personal behaviour
- Reduces egoism and self –centeredness by reconnecting mind, body, and soul
- Increases capacity for wisdom and compassion.

The goal of meditation is not to clear the mind from all thoughts, but to become aware of each thought as it arises, to acknowledge it without getting caught up in it, and then to return the attention to the focal point of the meditation.

Recreation and Inner Work through Self-Exploratory Means

Military leaders can increase their self-awareness by doing inner work through self-exploratory means such as reading, writing, and journaling of daily events by maintaining a diary,[52] all of which are known to contribute to positive outcomes.

Readers should be cautioned that, while reading is a good hobby, they must think critically about the information contained in self-help books or on the internet, because this information may be subjective and misleading.

Navigating Personal Issues for Military Leaders
Personal Issues Related to Individual Self-Awareness

The military expects its leaders to be intra-personally and interpersonally competent, knowledgeable, well informed, and self-aware. Military leaders are required to differentiate their personal moral code from and reconcile it with that of the professional military values. This clearly hinges on exploring and identifying personal values that underlie one's moral beliefs (for example, Hinduism in India[53] is based on the philosophy of nonviolence, yet Indian soldiers practicing the Hindu way of life may have to neutralize the enemy to prevent their own team members from being put in harm's way). While morals indicate right from wrong and good from bad and refer to one's beliefs, they also represent standards used as guidelines in the military context and are based on broader standards in the military. In contrast, values are less socially and more individually derived; they are the bedrock of beliefs and attitudes that provide direction in everyday life in the military. Leaders' decision- making process is based on five moral principles, namely autonomy, beneficence, non-maleficence, justice, and fidelity.

(a) Autonomy[54] refers to leaders' willingness to honor soldiers' right to make their own decisions (and to take ownership of those decisions). This is not to suggest that free will is a soldier's right, yet military leaders would do well to understand that it is more about free choice in that soldiers have the right to decide and choose their own life direction. It may include facilitating soldiers' self-determination, allowing soldiers to become the best version of themselves while ensuring that they avoid becoming dependent on their leaders. The first author[55] would like to share an incident which later made him realize that his own skills and enthusiasm in demonstrating weapon handling as an instructor may not have been the best way to teach at the weapons small arms firing range where there is no room for error during 'live firing.'

> I had just come back to the regiment after completing a platoon weapons course and was promptly detailed as the officer in charge of the Annual Range Classification Firing. When one of the soldier's weapons stopped firing during the course of training, he raised his right arm as per standard procedure to indicate

that there was a problem with his rifle. I immediately walked up to the squad and my newly acquired knowledge told me that the position of the gas regulator in the rifle needed to be changed for smooth operation. I promptly rectified the problem by changing the gas regulator position myself so that firing could resume smoothly. Little did I realize that in my exuberance to show off my weapon handling skill as a young officer, I had just created dependency for the soldier who may not find me around when his weapon stops firing during military operations with the enemy when he raises his right hand and looks over his shoulder to find nobody there to rectify the defect in weapon firing. Subsequently, I would make every soldier remove firing defects in their weapons themselves and this gave them more confidence in themselves, in their weapons and their leader who would trust them to treat their weapons with caution and care.

(b) Beneficence[56] in the military context refers to leaders' commitment to do well for soldiers under their command and seek to promote soldiers' competence (to increase their chance of survival during operations) through realistic training, ensure their welfare as well as their dignity and respect. Leaders need to pay individual attention to soldiers' different training and developmental needs in the personal and professional domains.

(c) Non-maleficence[57] implies that leaders hold no personal grudge or animosity and communicate clearly that they mean no harm. In other words, doing no harm refers to both the intention of no harm and the selection of purposeful behaviour that will not cause harm and takes into account individual and cultural differences among soldiers.

(d) Justice[58] refers to transparency in dealings with as well as fair and equal treatment of soldiers without discriminating on place of birth, caste, religion, creed, or gender. Leaders displaying distributional justice would ensure that all troops under their command have equal opportunities, equal access, equal treatment, and equal fairness in requests for leave, professional courses, promotion, transfers and postings as well as recreational and welfare facilities.

(e) Fidelity[59] for military leaders refers to the espoused military values of honesty, loyalty, and commitment to serve the organization (read as soldiers under your command). It also refers to the psychological contract in keeping promises and living up to the Chetwode code of conduct[60] for officers, a code that has been taught to them during their formative military training in cadet training institutions.

How leaders in the military view and apply these moral principles may be guided by their personal values, beliefs, attitudes, and interpretation and hence the need for leaders to have an insight into their individual level of self-awareness. Since values are enduring beliefs that guide personal preferences and decision-making, they reflect leaders' feelings or attitudes, which then translate into preferred actions and behaviors. Although values and morality influence each other, they are not the same. Some values may be driven by an evaluative perspective and may be moral (e.g., soldiers choosing a vegetarian diet for moral evaluative reasons if their choice is based on their belief that being vegetarian protects non-human lives and saves God's creatures. However, their choice is not based on a moral decision if it is made strictly on the belief that a vegetarian lifestyle is merely conducive to good health).

Individual self-awareness about values is critical since leaders cannot hide their values from their superiors and subordinates who have many opportunities to observe them. Indeed, the values manifest in many ways, including dressing style, non-verbal behavior, interpersonal relations and communication style across the chain of command. Only through self-awareness and self-exploration will leaders develop a better insight about how their values contribute to the way they exercise their command.

Personal Issues Related to Interpersonal Self-Awareness

Once leaders have achieved a good awareness about their own belief system and personal values, they can then turn their attention to the implicit choices that they make in interpersonal contexts. While interpersonal self-awareness relates to all five moral principles (i.e., autonomy, beneficence, non-maleficence, justice and fidelity), critical interpersonal dynamics include many other factors such as trust, commitment, integrity, approval from significant others, identification with others, social skills, emotional expressiveness, confidence in relating to others and ability to maintain healthy relationships. While these interpersonal skills are not exhaustive, it gives a starting point as well as an idea about what kinds of patterns and interactions military leaders need to explore. For example, leaders who are uncomfortable in the company of others may inadvertently communicate their awkwardness to their command /team members who, in turn, may take responsibility for their leaders' discomfort and may start to search for causes within themselves. Leaders with poor social skills may not be successful in connecting with or relating to their chain of command, not because of insufficient caring, or empathy, but because of mannerisms that may be perceived as offensive, or be uncomfortable or confusing to team members. Therefore, the first step in developing interpersonal self-awareness is for leaders to recognize that interpersonal relationships could be improved were they to change their own behavior. The second step is to identify how, for each interpersonal skill, leaders behave in their relationship with their command / team members. Leaders have then to evaluate how their pattern of behavior may affect their work relationships. Finally, leaders need to select which interpersonal behaviors to fine tune in order to improve their relationship with their team members.

Personal Issues Related to Cultural Self-Awareness

In its broadest term, culture[61] refers to any group that shares and transmits a set of values, beliefs and learned behaviors. As multicultural operations are now very common in the military environment across the globe, leaders must develop cultural self-awareness,[62] which requires them to examine their perceptions of and attitudes about other cultural groups and to evaluate their level of cultural sensitivity and competence through introspection and learning. Leaders may develop cultural sensitivity and competence through introspective work that requires much self-exploration and openness. Cultural awareness[63] refers to the process of recognizing personal biases, prejudicial beliefs, and stereotypical attitudes, behaviors, and reactions. Cultural awareness is gained through self-reflection and respect for others, as well as through the recognition and belief that differences are not synonymous with challenges and conflicts. Cultural knowledge can be accrued by exposure to and familiarization with different cultures. On the other hand, cultural skills[64] may be acquired and developed through learning about alternative approaches to reduce cultural prejudices, biases, and stereotyping.

Conclusions

An extensive review of the literature on enhancing functional fitness suggests that self –regulation models of behavior are concerned with what leaders choose to do and how they go about achieving missions, goals, and desired outcomes at the individual and team level. A better understanding of how to enhance functional fitness and to identify the gaps in leaders' behavior as well as the areas for improvement may help reduce leaders' skills deficits. Encouraging leaders to self-develop with relevant initiatives may lead not only to better leaders but also to increased performance and improved work relationships, thereby creating a positive work environment in the military.

More studies on the dimensions of self-regulation are needed as the association between these various dimensions remain poorly understood. In addition, the impact of self-regulation on functional fitness in the military remains elusive, although many studies have examined leadership and teamwork. Identifying the associations between leaders' functional fitness and their self-regulation skills may provide directions for future research.

Directions for Future Research

The focus of the chapter has been on highlighting and better understanding leaders' functional fitness competencies through self-regulation initiatives in the military. While a plethora of literature supports the view that leaders' effectiveness is directly proportional to their knowledge, skills and competencies, the critical specific intrapersonal competencies are not given the attention they deserve. Future research could focus on studying leaders' self-regulation competencies that may positively affect military well-being and health, which have a direct bearing on military effectiveness.

Hence, future studies could investigate military leaders on different dimensions of self-regulation and further examine dimensions relative to rank, element, occupation, and gender. Such studies may also result in training strategies to foster the development of intrapersonal skills in an effort to improve individual and collective health, thereby enhancing performance, despite operating in highly demanding military environments.

Notes

1 M. Parmak, "Health beyond symptoms: Multidimensional approach to functional fitness in Military Organisations." In Rawat (Ed.), *Military psychology – International perspectives*, pp. 147–160 (2017). Jaipur: Rawat Publications; S. Rawat, "Identifying malingering in cadets: Training capsule for newly posted instructors" conducted in June 2010 by Lt Col Dr Samir Rawat, Psychologist & Instructor Class 'A.' National Defence Academy, Khadakwasla, India; S. Rawat, "Social media challenges to the military and functional fitness among soldiers." Workshop at the Military Behavioural Sciences Conference organised by the Ministry of National Defence in Bucharest, Romania, June 2017; W.B. Jonas, F.G. O'Connor, P. Deuster, J. Peck, C. Shake and S.S. Frost, "Why total force fitness." *Military Medicine*, 175(8S) (2010), 6–13.
2 S. Rawat and A. Deshpande, "Complexity and leadership challenges in twenty-first century warfare." In Dobre, Cristian (Ed.), *Psicologia Militara- Provocari Ale Secolui XXI*, Vol. 4, pp. 46–55 (2019). Bucharest: Central Technic-Editorial Al Armatei, Bucharest.
3 S. Rawat, "Cognitive and volitional perspectives in enhancing human factors." In A. MacIntyre, D. Charbonneau and D. Lindsay (Eds), *Military performance*

in human factors in military performance (2019). Kingston: Canada Defence Academy Press, In Press.
4 M.J. Bates, J.J. Fallesen, W.S. Huey, G.A. Packard, D.M. Ryan, C.S. Burke and S.V. Bowles, "Total force fitness in units Part 1: Military demand-resource model." *Military Medicine*, 175, no. 8S (2013): 1164–1175.
5 F.C. Budd and S. Harvey, "Military fitness –for- duty evaluations." In C.H. Kennedy and E.A. Zillmer (Eds.), *Military psychology – Clinical and operational applications*, pp. 35–60 (2006). New York: The Guildford Press; K. Haggerty, "Changing lifestyles to improve health." *Preventive Medicine*, 6 (1977): 276–283; for an overview, see Parmak, 2017.
6 M. Parmak, "Health beyond symptoms".
7 WARMCAMP, an Acronym for Workshop and Roving Mezzanine Conference of Applied Military Psychology is an International Conference conducted annually in India.
8 G. Fischler, Assessing fitness- for- duty and return- to-work readiness for people with mental health problems. *The MCDA Communiqué* (2000). Apple Valley, MN: Minnesota Career Development Association.
9 Military Demand-Resource model highlights the dynamic interactions across demands, resources, and outcomes; The MDR model is a systems-based framework that identifies key demands and resources thought to influence psychological fitness end states. The MDR model includes four primary components that identify: (1) demands placed on the unit, (2) outcomes associated with a unit's fitness (e.g., performance and resilience), (3) resources that mitigate the impact of demands on the outcomes, and (4) feedback loops that account for interactions among factors and time, especially regarding balancing demands and resources. M.J. Bates, S.V. Bowles, J. Hammermeister, C. Stokes, E. Pinder, M. Moore and G. Burbelo, "Psychological fitness: A military demand-resource model." *Military Medicine*, 175, no. 8 (2010): 21–38.
10 R.M. Bray, J.L Spira, K. Rae Olmsted and J.J. Hout, "Behavioural and occupational fitness." *Military Medicine*, 157, no. 8S (2010): 39–56.
11 W.B. East, *A historical review and analysis of Army physical readiness training and assessment* (2013). Fort Leavenworth, KS: Combat Studies Institute Press.
12 G.P. Krueger, "Contemporary and future battlefields: Soldier stresses and performance." In P.A. Hancock and J.L. Szalma (Eds.), *Performance under stress*, pp. 19–44 (2008). Aldershot, UK: Ashgate.
13 C.S. Carver, & M.F. Scheier, "Self-regulation of action and affect." In K.D. Vohs and R.F. Baumeister (Eds.), *Handbook of self-regulation: Research, theory and applications*, 2nd ed., pp. 3–21 (2011). New York: Guilford Press.
14 C.S. Carver and M.F. Scheier, "Aspects of self, and the control of behavior." In B.R. Schlenker (Ed.), *The self and social life*, pp. 146–174 (1985). New York: McGraw-Hill.

15 R.G. Lord, J.M. Diefendorff, A.M. Schmidt and R.J. Hall, "Self-regulation at work." *Annual Review of Psychology*, 61 (2010), 543–68.
16 J.B.Vancouver and D.V. Day, "Industrial and organizational research on self-regulation: From constructs to applications." *Applied Psychology: An International Review*, 54 (2005): 155–85.
17 K.D. Vohs, R.F. Baumeister, B.J. Schmeichel, J.M. Twenge, N.M. Nelson, and D.M. Tice, "Making choices impairs subsequent self-control: A limited-resource account of decision making, self-regulation, and active initiative." *Journal of Personality and Social Psychology*, 94 (2008): 883–98.
18 M. Muraven and R.F. Baumeister, "Self-regulation and depletion of limited resources: Does self-control resemble a muscle?." *Psychological Bulletin*, 126 (2000): 247–259.
19 M. Muraven, D.M. Tice, and R.F. Baumeister, "Self-control as a limited resource: Regulatory depletion patterns." *Journal of Personality and Social Psychology*, 74 (1998): 774–789.
20 US Field Manual FM 6-22 Leader Development. (2015) Chapter 7: 113–176.
21 L.E. Atwater and F.J. Yammarino, "Does self-other agreement on leadership perception moderate the validity of leadership and performance predictions?" *Personnel Psychology*, 45 (1992): 141–164; L.E. Atwater and F.J. Yammarino, "Self-other rating agreement: A review and a model." *Research in Personnel and Human Resources Management*, 15 (1997): 12 1–174; A. Bandura. *Social learning theory* (1977). Englewood Cliffs: Prentice-Hall; A. Bandura. *Social foundations of thought and action: A social cognitive theory* (1986). Englewood Cliffs, NJ: Prentice Hall; A. Bandura, "Social cognitive theory of self-regulation." *Organizational Behavior and Human Decision Processes*, 50 (1991): 248–287; A. Bandura, Self efficacy: The exercise of control (1997). New York: H.W. Freeman; C.S. Carver & M.F. Scheier, "Aspects of self, and the control of behavior." In B. R. Schlenker (Ed.), *The self and social life*, pp. 146–174 (1985). New York: McGraw-Hill; A.K. Korman, Hypothesis of work behavior revisited and an extension. *Academy of Management Review*, I (1976): 50–63; A.S. Tsui, & S.J. Ashford, "Adaptive self-regulation: A process view of managerial effectiveness." *Journal of Management*, 20 (1994): 93–121.
22 L.E. Atwater, C. Ostroff, F.J. Yammarino, and J.W. Fleenor, "Self-other agreement: Does it really matter?" *Personnel Psychology*, 51 (1998): 577–598; R.F. Baumeister, T.F. Heatherton, and D.M. Tice, *Losing control: How and why people fail at self-regulation* (1994). San Diego, CA: Academic Press.
23 C.S. Carver and M.F. Scheier, "Outcome expectancy, locus of attribution for expectancy, and self-directed attention as determinants of evaluations and performance." *Journal of Experimental Social Psychology*, 18 (1982), 184–200.
24 J.Y. Shaw, "Automatic for the people: How representations of others implicitly affect goal pursuit." *Journal of Personality and Social Psychology*, 84 (2003): 661–681; J.Y. Shaw and A.W. Kruglanski, "When opportunity knocks: Bottom-up

priming of goals by means and its effects on self-regulation." *Journal of Personality and Social Psychology*, 84 (2003): 1109–1122.
25. B.J. Zimmerman, "Becoming a self-regulated learner: An overview." *Theory into Practice*, 41 (2002): 64–70.
26. L.E. Atwater, C. Ostroff, F.J. Yammarino, and J.W. Fleenor, "Self-other agreement: Does it really matter?" *Personnel Psychology*, 51 (1998): 577–598; R.F. Baumeister and K.D. Vohs, "Self-regulation, ego depletion, and motivation." *Social and Personality Psychology Compass*, 1 (2007): 115–128.
27. D. Goleman, *Working with emotional intelligence* (1998). New York: Bantam; C. Peterson and N. Park, "Character strengths in organisations." *Journal of Organizational Behavior*, 27 (2006): 1149–1154.
28. R.F. Baumeister, "Self-regulation, ego depletion, and inhibition." *Neuropsychologia*, 65 (2014): 313–319; C.C. Manz and H.P. Sims. *Business without bosses: How self-managing teams are building high performance companies* (1993). New York: Wiley.
29. M. Snyder and T.C. Monson, "Persons, situations, and the control of social behavior." *Journal of Personality and Social Psychology*, 32 (1975): 637–644; E.L. Tobey and G. Tunnell, "Predicting our impressions on others: Effects of public self-consciousness and acting, a self-monitoring sub-scale." *Personality and Social Bulletin*, 7 (1981): 66, 1–669.
30. R.E. Levasseur, "People skill: Self-awareness-a critical skill for MSlOR professionals." *Interfaces*, 21 (1991): 130–133; J. D. Mayer and P. Salovey, "Emotional intelligence and the construction and regulation of feelings." *Applied & Preventing Psychology*, 4 (1995): 197–208.
31. R.F. Baumeister, "The self." In D. T. Gilbert, S. T. Fiske, and G. Lindzey (Eds.), *Handbook of social psychology*, 4th ed., Vol. 2, pp. 680–740 (1998). New York: McGraw-Hill; S. Rawat, A.P. Deshpande, and G. Singh, "Self perception as stress buffering in military." In A. Pesic (Ed.), *Stress in the military profession*, Vol. 3, pp. 65–80 (2018). Strategic Research Institute, Ministry of Defence, Republic of Serbia.
32. C.S. Carver and M.F. Scheier, "Aspects of self, and the control of behavior." In B. R. Schlenker (Ed.), *The self and social life*, pp. 146–174 (1985). New York: McGraw-Hill; C.S. Carver and M.F. Scheier, "Principles of self-regulation." In E. T. Higgins and R. N. Sorrentino (Eds.), *Handbook of motivation and cognition*, Vol. 2, pp. 3–52 (1990). New York: Guilford.
33. J.S. Bunderson and K.M. Sutcliffe, "Management team learning orientation and business unit performance." *Journal of Applied Psychology*, 88, no. 3 (2003): 552–560.
34. G.P. Latham and C.C. Pinder, "Work motivation theory and research at the dawn of the twenty-first century." *Annual Review of Psychology*, 56 (2005): 485–516.

35 P.G. Northouse, *Leadership*, 4th ed. (2007). Thousand Oaks: Sage Publications; G. Yukl. *Leadership in organisations*, 3rd ed. (1994). Englewood Cliffs, NJ.: Prentice Hall.
36 A. Bandura, *Self-efficacy: The exercise of control* (1997). New York: H.W. Freeman; C. Peterson & N. Park, "Character strengths in organisations." *Journal of Organizational Behavior*, 27 (2006): 1149–1154; S. Rawat, "Cognitive and volitional perspectives in enhancing human factors: In military performance." In A. MacIntyre, D. Charbonneau and D. Lindsay (Eds.), *Human factors in military performance* (2018). Kingston: Canadian Defence Academy Press; W. Mischel, N. Cantor and S. Feldman, "Principles of self-regulation: The nature of willpower and self-control." In E.T. Higgins & A.W. Kruglanski (Eds.), *Social psychology: Handbook of basic principles*, pp. 329–360 (1996). New York: Guilford.
37 P.C. Early, T. Connolly, and T. Ekegren, "Goals, strategy development and task performance: Some limits on the efficacy of goal setting." *Journal of Applied Psychology*, 74 (1989): 24–33.
38 M. LaLonde, *A new perspective on health of Canadians – A working document* (1974). Ottawa Information Canada.
39 J. Arnold, I.T. Robertson, and C.L. Cooper, *Work psychology: Understanding human behaviour in the work place*, Vol. 13, pp. 226–245 (1996). New Delhi: Macmillan India Limited; B. Avolio, W. Gardner, F. Walumbwa, F. Luthans, and D. May, "Unlocking the mask: A look at the process by which authentic leader's impact follower attitudes and behaviors." *The Leadership Quarterly*, 15, no. 6 (2004): 801–823.
40 A. Antonovsky, *Health, stress and coping* (1979). San Francisco: Jossey Bass.
41 A. Bandura, *Self efficacy: The exercise of control* (1997). New York: H.W. Freeman; K.D. Vohs and R.F. Baumeister. Understanding self-regulation: An introduction. In R.F. Baumeister & K.D. Vohs (Eds.), *Handbook of self-regulation: Research, theory, and applications* (2004). New York: Guilford Press.
42 Ibid.
43 Ibid.
44 C.S. Carver and M.F. Scheier, "Principles of self-regulation." In E.T. Higgins and R.N. Sorrentino (Eds.), *Handbook of motivation and cognition*, Vol. 2 (1990). New York: Guilford.
45 J.B. Vancouver, "Integrating Self-regulation theories of work motivation into a dynamic process theory." *Human Resource Management Review*, 18 (2008): 1–18; S. Rawat Holistic health and well being for optimal officer cadet development, In Lt Col Dr Samir Rawat (Ed), *Cadet diary psychology of warrior ethos and cadet leadership development*, Vol. 3, pp. 63–76 (2018). Jaipur, India: Rawat Publications.
46 L.S. Das, *Awakening of the sacred: Creating a spiritual life from scratch* (1999). New York: Broadway Books.

47 D. Greenberger and C.A. Padesky, *Mind over mood: Change how you feel by changing the way you think*, 2nd ed., p. 57 (2016). New York: Guilford Press. Tim Bowden and Sandra Bowden, "Acceptance and Commitment Therapy (ACT): An overview for practitioners." *Australian Journal of Guidance and Counselling*, 22, no. 2 (2012): 279–285.
48 A. Fogelsanger, *See yourself well. Guided visualisation and relaxation techniques* (1994). Brooklyn Equinox; M. Rossman. Mind/Body Medicine, *How to use your mind for better health* (1993). New York: Consumer Report Books.
49 Ibid.
50 L. LeShan, *How to meditate* (1974). New York: Bantam.
51 L.S. Das, *Awakening of the sacred*.
52 J.W. Pennebaker, *Opening up. The healing power of expressing emotions* (1990). New York: Guildford.
53 Hinduism is not a religion, but a way of life as prescribed in Indian scriptures and ancient texts from 5000-year-old Indian civilization.
54 D.G. Ancona and D.F. Caldwell, "Bridging the boundary: External activity and performance in organizational teams." *Administrative Science Quarterly*, 37 (1992): 634–655; B.J. Avolio and W.L. Gardner, "Authentic leadership development: Getting to the root of positive forms of leadership." *The Leadership Quarterly*, 16, no. 3 (2005): 315–338; S. Rawat, "Developing good military habits: Insights for cadets from a combat veteran and military psychologist." In Lt Col Dr Samir Rawat (Ed.), *Cadet diary 2.0 psychology of warrior ethos and cadet leadership development*, Vol. 1., pp. 25–36 (2019). Jaipur, India: Rawat Publications.
55 Lt Col Dr Samir Rawat, is a combat veteran and military psychologist from India.
56 Barbara Adams, David Bryant, & Robert Webb. *Trust in teams literature review* (2001). Canadian Department of National Defence, DCIEM No CR-2002-042; B.M. Bass & B.J. Avolio, "Transformational leadership: A response to critiques." In M. M. Chemers and R. Ayman (Eds.), *Leadership theory and research: Perspectives and direction*, pp. 49–80 (1993). San Diego: Academic Press.
57 C. Brems, "Dimensionality of empathy and its correlates." *Journal of Psychology*, 123 (1989): 329–337; S. Rawat. *Holistic health and well being*.
58 S. Rawat, "Transitioning from a cadet to a professional officer: A psychological perspective to leadership traits and competencies." In S. Rawat (Ed.), *Cadet diary – Psychology of Warrior Ethos and leadership development* (2018). Jaipur, India: Rawat Publications.
59 S. Rawat, "Communicative behavior as transformative power of military leadership." In D. Watola and A. MacIntyre (Eds), *From knowing to doing: International perspectives on leadership effectiveness* (2018). Kingston: Canadian Defence Academy Press.

60 Chetwode Code – "The safety, honour and welfare of your country come first, always and every time. The honour, welfare, and comfort of the men you command come next. Your own ease, comfort and safety come last, always and every time." This is known as the "Chetwode Motto" and is the motto of the officers passing out from the Indian Military Academy.
61 Allan D. English, *Understanding military culture: A Canadian perspective*, p. 15 (2004). Montreal and Kingston: McGill-Queen's University Press; Karen D. Davis and Brian McKee, "Culture in the Canadian forces: Issues and challenges for institutional leaders." In Robert W Walker (Ed.), *Institutional leadership in the Canadian forces: Contemporary issues* (2007). Kingston: Canadian Defence Academy Press.
62 T.M. Singelis, *Teaching about culture, ethnicity and diversity: Exercises and planned activities* (1998). Thousand Oaks, CA: Sage.
63 T.M. Singelis and W.J. Brown, "Culture, self and collectivist communications. Linking culture to individual behaviour." *Human Communication Research*, 21 (1995): 354–359.
64 S. Rawat and Abhijit Prakash Deshpande, "Relevance of self regulation for military leaders." In Lt Col Dr Samir Rawat (Ed.), *Cadet diary 2.0 psychology of Warrior Ethos and cadet leadership development*, Vol. 2., pp. 37–54 (2019). Jaipur, India: Rawat Publications.

Madlaina Niederhauser and Hubert Annen

Resilience and Resilience Training: Focus on Military Science

> *This chapter focuses on positive psychology, resilience and resilience training with the main emphasis on military. After a brief introduction to stress, the two salutogenetic concepts "positive psychology" and "resilience" were reviewed. Further, resilience factors will be explained and different resilience training methods and their effects will be identified. Finally, a discussion about further studies and practical implications for military organizations concludes this chapter.*

What about Stress?

Nobody can escape the adversities of life. Therefore, the phenomenon of stress receives great attention in lay conception as well as in psychological science. Triggers of stress (stressors) occur in various ways (daily stressors or critical life events) and frequencies (acute or chronic).
[1] Whether a situations are interpreted as stressful or not varies from one individual to another. By cognitively evaluating the stressful situation, a person assesses whether there is a potential risk to themselves and to their health. This process consists of a primary and secondary appraisal. The primary appraisal assesses whether the specific situation represents a genuine threat or not. If the situation is assessed as a threat, the secondary appraisal follows, in which available resources are evaluated to reduce or avoid the threat. Stress occurs when the situation cannot be managed with the available resources.[2] Therefore, stress is defined as a subjectively perceived imbalance between demands and reaction possibilities or as an individual reaction to environmental demands that are difficult or impossible to cope with.[3]

As soon as the situation is evaluated as a source of stress, a physiological and cognitive-emotional stress reaction takes place. The unspecific reactions involved when one experiences stress provide the necessary energy and cognitive processes to adapt to the change and are summarized by Selye as a general adaptation syndrome.[4] First, the body is set in alarm mode and mobilizes the necessary energy. In the resistance stage, the body adapts to the sustained threat. However, this stage can only be maintained for a limited time. If the stressful situation continues, the stage of exhaustion follows.[5] Prolonged stress can lead to negative physical and psychological effects.

Physiological and Psychological Effects of Stress

Consequences of stress can be short-, long-term or both and even single acute stressors can have an immediate influence on the physiological stress reaction of our body. However, chronic stress results in exhaustion of an organism and can lead to the development of physical disease[6] like chronic fatigue, cardiovascular disease, diabetes mellitus, gastrointestinal disorders or chronic pain.[7] Further, psychological effects like an increasing risk of developing depression disorders[8] or post-traumatic stress disorders (PTSD) because of trauma are possible.[9]

Stress in Military Life

The phenomenon of stress is also a major topic for military psychology, because stress is inherent in everyday military life.[10] Everyday military stressors are the main contributors, because, at any given time, only a minority of military personnel participates in actual combat.[11] From the start of a military member's career, basic military training is immediately life changing. Recruits undergo a series of physical and psychological challenges during initial training[12] and the transition from civilian to military life is often accompanied by a sense of loss, disillusionment, and disappointment.[13] Basic military training involves highly structured daily activities and precludes personal autonomy.[14] Recruits lose personal control over their lives and must quickly learn new standards of behavior. Earlier sources of support such as friends and families are no longer directly available. In addition, coping strategies previously learned tend to be ineffective.[15] Furthermore, long working hours, time pressure and limited personal freedom make every day military life a challenge not only during basic military training, but thereafter.[16] These stressors increase the likelihood of developing mental and physical health problems.[17] For instance, the most commonly reported stress-related problems by military personnel include aggression, excessive alcohol consumption, fatigue, and sleep disorders.[18]

Psychological stress can also have adverse psychosocial consequences for deployed military personnel. Traumata during combat exposure can lead to pathological responses like PTSD. Studies estimate that approximately 20 % of deployed soldiers with combat exposure develop PTSD and related symptoms.[19]

To sum up, military service is highly stressful. It is understandable that psychological difficulties and challenges will arise frequently and are created by a wide range of uncertainties, urgencies, novelties, and risks that characterize military life.[20] Furthermore, psychological stress factors vary from one phase of a member's military career to the next[21] and all the above situations can result in extensive emotional and behavioral consequences. Therefore, it should be the goal of military organizations to prevent or minimize performance losses due to chronic stress effects and stress-related diseases.

A Salutogenetic Approach

Fortunately, people are mostly able to deal with stressful situations. In a study with Israeli citizens, exposed to the Palestinian Intifada, most of them were able to adapt to the situation and stay mentally healthy despite the permanent distress.[22] Similar results were observed in children who grow up in critical circumstances. In a longitudinal study, these children were observed and studied for over 40 years. Nearly one third of the children involved in the study developed into healthy adults despite their difficult circumstances.[23] Thus, the focus moved away from the pathogenetic to the salutogenetic approach, which investigates what keeps people healthy[24] and what makes life most worth living.[25] Subsequently, two areas of this salutogenetic approach, positive psychology and resilience will be presented in more detail.

Positive Psychology

Since the Second World War, psychology was dominated by a disease model of human behavior and the field of psychological research had been limited to the cure of mental illness. Martin Seligman wanted to change this view and introduced positive psychology in 1998 during his term as president of the American Psychologists Association (APA). Positive psychology is an umbrella term for theories and research about what makes life most worth living, e.g., happiness, well-being, satisfaction, virtues, character strengths and talents.[26]

Positive psychology consists of three pillars: positive emotions, positive character, and positive institutions. Some typical examples of positive emotions include joy and hope. These can be differentiated by time: past-oriented emotions such as contentment, satisfaction and pride, future-oriented emotions such as optimism and trust or present-oriented emotions such as pleasure and gratifications. A positive character refers, for example to friendliness, humor, and strength of character. Positive institutions describe the framework conditions of institutions (e.g., family, school, or workplace) for the health and growth of the persons belonging to them.[27]

Interventions of Positive Psychology

Interventions from positive psychology are intentional activities or treatment methods to cultivate positive feelings, behaviors, and cognitions.[28] The aims are simple intervention methods that are suitable for everyday use, which support people in using their strengths, making full use of resources and potential and thus leading to a fulfilled life even when dealing with crises.[29] A meta-analysis has demonstrated that positive interventions can help to increase well-being and life satisfaction and reduce depressive symptoms.[30]

Positive Psychology in Military Organizations

In 2008, positive psychology was introduced in military organizations. This was not seen as a replacement for traditional models or methods, but rather as a supplement to the military psychologist's toolbox.[31] The practical implementation of positive psychology in the military can generate added value, especially in the areas of leadership, training, and education. For instance, it can be applied by high-quality selection processes based on strengths, creating a motivating environment with positive experiences through appropriate leadership styles, or the promotion of individual self-reflection and resilience.[32]

It is evident that positive psychology is related to well-being and life satisfaction. In stressful times, as the military context often is, more intervention tools are needed than those provided by positive psychology. Therefore, the following sections of this chapter will take a closer look at the topic of resilience and illustrate in some places where positive psychology is reflected in this construct and how intervention techniques have been adopted from positive psychology.

Resilience

Psychological resilience refers to the phenomenon that many people who have experienced psychological or physical suffering do not, or only temporarily, become ill.[33] Since the turn of the millennium, interest in resilience in the social sciences has risen sharply. However, despite the large amount of research, there is no universally accepted definition of psychological resilience.[34]

Definition of Resilience

Over the past two decades, the concept of resilience has significantly changed from a trait- to a process-oriented approach. The earlier approach defined resilience as personal trait or a set of characteristics of an individual that provides protection from the negative effects of stress and trauma. Resilience conceptualized as a trait is considered as being relatively stable over time.[35] However, there is only weak empirical evidence for this assumption that resilience is a stable aspect of personality.[36]

The process-oriented approach views resilience as a dynamic process through which positive outcomes are achieved in the context of adversity.[37] The American Psychological Association defines resilience as *"the process of adapting well in the face of adversity, trauma, tragedy, threats or significant sources of stress."*[38] Therefore, resilience is the process of successfully adapting to stress or trauma, which is possible due to various protective factors.[39]

It is well established, that resilience is the successful adaptation to adversity. However, there is a disagreement regarding the meaning of a successful adaptation. For some authors, successful adaptation means maintaining normal functions, others define it as bouncing back from adversity.[40] These differences in conceptualization are most likely derived from the context in which resilience is studied. In the context of daily stressors, maintaining sustained functioning captures resilience better, while after a traumatic event bouncing back is more appropriate to describe someone as resilient.[41]

Resilience is also not a simple, static construct that relates equally to all stressors. It is possible that someone can adapt well to stress at work but cannot cope with stress in their personal life and/or relationships,[42] or is resilient towards one type of trauma but not another.[43] Furthermore, one's resilience level can change over time because of the interactions between environment and experienced stress and trauma.[44]

Resilience can only be determined if a person is or has been exposed to stress or trauma.[45] Therefore, a new approach defines resilience as an outcome and focuses on the many behavioral, cognitive, and affective components of resilience that condition the ability to survive well in adverse circumstances. From this perspective, resilience is conceptualized as a net effect.[46,47]

Factors of Resilience

Resilience consists of various factors, such as capabilities and resources, which are available to protect an individual from negative consequences associated with adverse experiences.[48] These protective factors influence how a person reacts to adversity or trauma.[49] Resilience factors can be present at different levels, like internal and external protective factors and factors at the community level.

The range of internal, individual resilience factors is very large. For example, stable factors such as extraversion,[50] intellect and self-efficacy,[51] as well as self-esteem, optimism, mastery and hardiness[52] are associated with resilience. Additional internal attributes include hope,[53] coping,[54] positive affect[6] and self-regulation.[55] A meta-analysis examining a wide range of resilience factors concluded that the internal factors with the strongest evidence in the literature compromise positive coping, positive thinking, positive affect, realism, behavioral control and physical fitness.[56] The strong relationship of resilience with positive coping, positive thinking and positive affect point out well the connection to positive psychology. Especially positive thinking, which includes optimism, humor, hope and positive emotions, strongly reflects the basic idea of positive psychology.

External resilience factors include resources and supportive relationships,[57] particularly at the family level. On one hand, they pertain to emotional ties, like emotional bonding to family members which includes shared leisure time and recreation, and emotional, instrumental, informational, and spiritual support. On the other hand, good communication, closeness, intimacy, and attachment, nurturing and adaptability of the family contributes as well to resilience. Among the external factors, family support is the one that has received the most evidence.[58]

Protective factors at the community level include peer and nonfamily member relationships, nonfamily member social support and religion.[59] Protective factors at community level can be allocated to one of these factors: Belongingness, cohesion, connectedness, and collective efficacy. The community-level factor with the strongest evidence for promoting in the literature was belongingness.[60]

To sum up, resilience requires certain attributes, characteristics, and skills. First, a certain degree of intelligence and cognitive skills are needed to identify the heart of a problem and to plan relevant steps toward a solution. Second, optimism and trust in one's own skills and abilities (like self-esteem, self-efficacy) contribute also to these steps. Third, to deal effectively with the intense emotions that arise in stressful situations, healthy emotion regulation strategies must be available. Finally, overcoming a challenging situation the presence of social support is often a necessity. Hence, the affected person must have social resources, good communication and social skills to activate the individual social support.[61]

Measurement of Resilience

In order to obtain enhanced knowledge about resilience, for scientific investigations and for derive targeted trainings interventions, it is important to be able to measure resilience reliably and validly.[62] Yet, the measurement of resilience is as varied as the definitions and there is no gold standard. Most of the existing resilience scales measure resilience either as a trait or focus on the measurement of resilience-related constructs (e.g., self-efficacy).[63] A methodological review[64] of resilience measurement scales has shown that the following three scales have the best psychometric properties: The Brief Resilience Scale,[65] the Connor-Davidson Scale,[66] and the Resilience Scale for Adults.[67] With regard to the military context, the following two scales are also recommended:[68]

- Dispositional Resilience Scale:[69] This 15-item instrument measures the construct hardiness, which is related to resilience. The scale includes three distinct factors: Sense of control of one's life, commitment in terms of the meaning ascribed to new experience and openness to viewing change as a challenge.
- Response to Stressful Experience Scale:[70] This 22-item instrument is based on samples drawn from military units in deployment and reflect five resilience-promoting factors: meaning making and restoration, active coping, cognitive flexibility, spirituality and self-efficacy.

Further research needs to focus on the development of additional valid measures of resilience, especially for specific populations like military personnel.

Resilience in Military Organizations

Resilience is a central concept for military organizations when they are about to optimize the human performance. It is an important component in keeping leaders, military personnel, and military families healthy and fit for service,[71] despite increased occupational risks.[72] Indeed, the large number of stressors and adversities associated with military service as described above highlights the importance of resilience in military settings.

In addition to the earlier mentioned resilience factors, in military settings other factors must be considered as well. First, a positive command climate is important because it simplifies and promotes intra-unit interactions, positive role modelling, pride- and support building for the mission, and implementation of institutional policies. Second, teamwork is crucial for the resilience of military personnel. Third, cohesion or the ability of the unit to carry out combined actions and to connect members with each other is closely related to resilience. Of these factors, positive command climate has received the most evidence to promote resilience.[73]

Resilience is also important at the military recruit's' level. Previous research has shown that highly resilient soldiers left (quit or failed) basic military training less frequently than those reporting low resilience levels.[74] For the Royal Marines in the UK, resilience was a strong predictor of the successful completion of basic military training.[75] Soldiers reporting high levels of resilience also exhibited superior military capabilities and tended to opt more often for a military career than less resilient soldiers .[76] In the U.S. Army, it was also shown that the rank achieved in the military correlated significantly with resilience.[77]

Resilience is also important for officers. In chronically stressful situations, resilient officers reported being less strained, tense, and worried, and considered challenges to be less stressful. In extreme military survival training, resilient officer cadets felt mentally and physically better prepared and were more confident to successfully cope with such stressful circumstances. Retrospectively, they were less stressed, cultivated a constructive style of communication in conflict situations and felt mentally and physically more efficient than less resilient officer cadets. Further, in acute stress situations, resilient military personnel reported less stress and faster recovery after a stressful event,[78] and more positive and less negative affect before and after the event.[79]

Resilience seems to be helpful for military personnel in different phases of their career.[80] The added value of existing resilience for military personnel can be seen in a variety of life stages or service situations, ranging from simple mission preparation to situations where pure life support can be at stake.[81] For example, in the pre-deployment phase, self-efficacy, provision of information, and leadership were the most salient resilience factors. During deployment, group cohesion had the strongest effect on resilience, but self-efficacy, support provided by the home front and leadership were also significant predictors.[82]

One existing theory suggest that resilience is most relevant among people at high risk of experiencing stress or trauma,[83] like those encountered in military settings. Not only can knowledge about resilience be used by military organizations to identify groups of individuals at high risk of developing poor psychological health in the long term but also to prepare military personnel better for their service at home or abroad. These preparations can be accomplished through the development and conduct of specific resilience training with focus on reduction of mental health impairments caused by stressors.[84]

Resilience Can be Trained

Current thinking recognizes resilience as a skill that can be developed or improved and many training programs to promote resilience have been developed.[85,86]

The salutogenetic approach serves as an ideological orientation in the development of resilience training and thus the promotion of resilience interventions is predominantly preventive in nature.[87] Theoretically, the interventions are often based on the transactional stress model,[88] on the basics of cognitive behavioral therapy[89,90] and on positive psychology.[91] The interventions focus in particular on psychological aspects, like cognitive restructuring or recognizing and dealing with irrational thoughts and thinking traps, as well as on individual resources and character strengths. In addition, the physiological effects of stress can be attenuated through relaxation and reduction of physical arousal.[92,93,94] Furthermore, the training includes the psychosocial factors that are connected to resilience. Most frequently, there are training modules discussing not only self-efficacy, optimism, social resources, coping, but also positive communication and stress management, emotion regulation, energy management, problem solving, empathy and attachment.[95,96,97,98]

As mentioned above, a lot of resilience training programs included evidence-based methods from positive psychology. The idea is to supplement the programs with techniques, which increase positive emotions, optimism, and well-being, as well as the use of individual character strengths in daily life.[99]

The effects of resilience training have been evaluated using different clinical and non-clinical populations, and different formats, durations, and settings. A wide range of symptoms and maladaptive behavioral outcomes have been examined.[100,101] The remainder of the chapter gives an overview of the existing research on effects of resilience training in civilian and military contexts.

Resilience Training in Civil Life

Several studies have examined the influence of resilience training on psychological factors and performance in a working environment. Research findings include significant positive effects on general mental health,[102] psychological capital,[103] and resilience.[104] Additionally, some study results have demonstrated a reduction in harmful outcomes such as negative attribution style,[105] social strain,[106] stress,[107,108] and depression symptoms.[109,110] Furthermore, resilience training has been linked to improvement of mastery, positive emotions, personal growth, mindfulness, acceptance and autonomy.[111] With respect to work performance,[112] propensity to innovate, motivation,[113] and job satisfaction[114] increased. However, it is important to acknowledge that some studies found no effects.[115,116]

Other studies focused on the effects of resilience trainings for police officers and evaluated the effects of trainings in the context of realistic simulated stressful situations. Results revealed significant reductions in negative mood,[117] self-reported stress and negative emotions, as well as in symptoms of depression.[118] Improvements in family relationships, communication and cooperation within work teams were also noted.[119] With regard to the physiological effects, the studies found a lower heart rate reactivity and an increase in antithrombin.[120] These improvements in performance were demonstrated in comparison to a control group.[121,122]

Another well evaluated resilience training program is the SMART-OP (a computer-based stress management training), which was developed and tested for NASA. Participants in the training program reported significantly lower stress levels and more control over stressful situations after the training compared to a control group. In addition, the participants of the SMART-OP group recovered faster from a standardized stress situation than a passive learning group.[123]

Probably, the resilience training program that has received the most empirical attention in the civilian context is the Penn Resilience Program (PRP).[124] This cognitive-behavioral group intervention aims at preventing depression in 10- to 14-year-old students. With basic guidance from positive psychology, among other techniques, the PRP especially promotes optimism and assertiveness.[125] A meta-analysis investigating the effects of PRP supported its positive results. Indeed, PRP reduces and prevents symptoms of depression and offers long-lasting effects. The program reduces hopelessness and raises optimism and works equally well for children of different racial and ethnic backgrounds.[126]

Resilience Training in Military Settings

A growing number of strategies and programs are available to encourage and support resilience for military members.[127] Not surprisingly, many military organizations developed their own effective and sustainable resilience-strengthening programs to reduce the risk of mental health issues in their personnel. Some programs focus on prevention, others are used to prevent and/or reduce post-traumatic stress disorder after deployment.

Training for Military Personnel Undergoing Deployment

Mental health problems are common following military-related traumatic events experienced during deployment.[128] Therefore, some studies evaluated the effects of resilience-based interventions with this at-risk population, especially because of the ineffectiveness of single debriefing sessions.[129]

In particular, two studies investigated interventions with soldiers following a deployment. The first, a longitudinal study, examined three post deployment interventions (battlemind briefing, battlemind training, and stress education) and their effects on mental health. The results show that participants in the battlemind briefing intervention, who had a high combat exposure, reported fewer post-traumatic stress symptoms, sleep problems and depression symptoms than those in the stress education group. Similarly, participants in the battlemind training intervention reported fewer post-traumatic stress symptoms and fewer depression symptoms than those in the stress education group.[130] Further, the effects of battlemind training remained significant at the four-month follow-up in that the study participants reported fewer post-traumatic stress and depression symptoms as well as a higher life satisfaction compared to the control group participants. The second study demonstrated that improvements in attitudes about seeking mental health care emerged immediately after the training, but not at follow-up.[131] These findings demonstrate, that brief early interventions with at-risk groups have the potential to be effective at combatting stress reactions.[132]

Two other studies focused on interventions with veterans suffering from PTSD. The first explored the effects of a computer-based self-managed cognitive behavior therapy and found a decline in daily ratings of post-traumatic stress and depression symptoms. The participants who completed the full self-managed therapy program obtained lower PTSD, depression, and anxiety scores at six-month follow-up. One third of those achieved a high-end state functioning.[133] Another study involving PTSD patients found improvements in affective symptoms, positive emotional health, memory, and executive functions at posttest in the resilience-oriented intervention group compared to the control group.[134] Taken together, these findings suggest that both computer-based and group-sessions resilience-based interventions can have a positive impact on the psychological health of veterans.

Prevention-Oriented Training for Military Personnel

Numerous studies in different countries examined the effect of resilience training programs for soldiers. Some studies aimed at reducing various risk factors through targeted training. For example, an anger management training program for active-duty military personnel significantly reduced symptoms of anger.[135] Another training approach focused on coping efforts for military trainees to reduce attrition. But these 2 hours of stress management training achieved no effect in post measurement.[136]

Other military organizations aspire to promote the resilience of their personnel by offering training in many resilience techniques. For instance, mental health training sessions describing various strategies for coping, for increasing a sense of belonging, for decreasing thought distortions and for improving stress management were provided to naval recruits. Results show that the intervention group reported better group cohesion, problem-solving coping strategies, and perceived social support as well as fewer anger expression coping strategies. These effects were particularly noticeable for recruits at-risk for depression.[137,138] Another study evaluated the efficacy of a brief cognitive-behavioral program targeting causal attributions, expectancy of control, coping strategies, and psychological adjustment in a sample of Australian army recruits in basic training. Compared with an active control group, the intervention group participants reported more temporary and specific attributions, less reliance on self-blame coping, greater positive states of mind, and lower psychological distress.[139] The evaluation of a resilience training on US soldier's well-being and attitudes showed further positive results. The training was provided during the first few days of basic combat training. Participants of the intervention group had experienced a faster decrease in anxiety symptoms and greater confidence in helping others and received more positive ratings than the control group. However, the intervention failed to improve depression symptoms and sleep.[140]

The two most recently published studies showed different effects. One study explored the effects of a resilience-based intervention delivered by military trainers for UK Royal Air Force recruits. Topics covered in the training included capitalizing personal strength, emotion regulation, enhancing awareness of psychological symptoms and learning methods to promote resilience. The study failed to find any impact of this training on mental health and well-being, stigmatization, mental illness, alcohol use, cohesion, and perceptions of leadership.[141] The second study investigated a short coping and emotion regulatory self-reflection training for officer cadets and revealed positive effects on mental health. The self-reflection group, compared with a coping skill group, reported less depression and anxiety symptoms, as well as less perceived stressor frequency.[142]

In a comprehensive research project, the Swiss military has developed an army resilience training (ART) program and evaluated this in a longitudinal control group design using subjective and biological variables. Preliminary results are promising. Participants reported being able to apply the newly learned techniques in both military and civilian everyday life. Similarly, in a military survival training, more than half of the participants reported using the techniques. Furthermore, a majority of the participants stated that in conflict situations, they can communicate more constructively, cope better with negative events, and control their emotions more appropriately than before the training.[143] The training also seems to positively influence dispositional factors. While the intervention group showed a significant decrease in stress reactivity, this did not change in the control group.[144] After the training, the intervention group also had a significant increase in resilience in contrast to the control group.[145] The data analyses from this study are ongoing and it is expected that the results will be published soon.

Comprehensive Soldier and Family Fitness

The largest existing preventive resilience program comes from the U.S. Army. The Comprehensive Soldier and Family Fitness (CSF2) was launched[146] in 2009 and has the goal to create an Army that is just as psychologically fit as it is physically.[147] This is the first attempt to improve the psychological fitness of all members of a military organization including soldiers, family members, and civilian employees.[148,149]

The CSF2 program consists of the following elements: online self-assessment to identify relevant resilience aspects, online self-help modules tailored to the self-assessment, master resilience training and mandatory resilience training for army leaders.[150] The core element of the CSF program is the Master Resilience Training (MRT), which is based on the Penn Resilience Program. The influence of positive psychology is evident. The self-assessment focuses on strengths rather than weaknesses and consists of four elements: emotional, family, social and spiritual fitness. Further, the training modules are conceptualized to increase self-reflection and to strengthen positive things in life. In the emotional fitness module, for example, participants learn to amplify positive emotions and how to recognize when negative ones are out of control. Also, the MRT is based on concepts from positive psychology to build mental toughness, signature strengths and strong relationships.[151]

The CSF program has been criticized for being released without pilot testing and disregard of ethical principles such as proof of usefulness and declaration of consent.[152] So far, most of the results of the CSF program have been published in industry reports and not in peer-reviewed journals. A study showed self-reported utility of the newly acquired resilience competencies in their military jobs, in their civilian life, and during counselling sessions with soldiers. Self-reported improved competencies were further related to individual well-being.[153] Units that received MRT found 60 % fewer diagnoses of drug and alcohol abuse and 13 % fewer diagnoses of anxiety, depression and PTSD compared to units without

MRT.[154] The train-the-trainer approach also seems to be effective. Soldiers, who worked for leaders with MRT training, had higher values in emotional fitness, character, positive coping, friendship, and less catastrophizing in the Army Global Assessment Tool (GAT) than soldiers of leaders without MRT training.[155]

Effects of Resilience Training Programs

The conceptual heterogeneity of resilience hinders scientific evaluation and generalizability of results. A meta-analysis examined the short- and long-term effects of various training approaches. Resilience-building programs had the strongest proximal effect on improving performance, enhancing well-being, and preventing psychosocial deficits. The only statistically significant distal effect was found for preventing psychosocial deficits. Furthermore, the effects diminished substantially over time with an exception for those at greater risk of experiencing stress or who lack protective resources, for whom distally measured effects were stronger than proximal effects.[156]

Not only does the delivery format impact the effect size of the results, but so does experimental design. The most effective programs were those with a one-on-one delivery format, followed by classroom-based group delivery format. Studies employing single-group within-participants design report stronger effects than studies utilizing between-participant designs. Similarly, targeted programs (specific for participants) report weaker effects than universal programs. There is no difference between military and non-military samples.[157]

To sum up, the overall effect of resilience training programs is quite small.[158] But even small effects when applied to many individuals may result in significant benefits[159] for military organizations.

What's Next?

Forty years of scientific research about resilience answered many questions. Still, further studies are needed to improve our understanding of how resilience works in different situations and how resilience training programs can be conducted with more long-lasting effects, especially in military organizations. To encourage research, we are proposing possible research design, moderator variables and outcome measurements.

Research Design

To optimize the effectiveness of resilience-building programs, developers should assess the needs of their participants, because different types of protective factors may be relevant depending on the participants' stressful situations.[160] For example, in the military, contrary to civilian life, problem-focused coping is sometimes counter-productive because of the low autonomy work environment, which could lead to an increase in distress symptoms.[161]

Most of the resilience training programs reported in the literature were implemented in classroom settings.[162] This type of training is expensive and rather impractical.[163] Further studies should concentrate more on train-the-trainer approach, as this is more realistic for a military organization. A report on the effectiveness of MRT has shown that this approach can have desired effects[164] and from management studies there are indications that there are contagion effects (people are learning from trained people).[165] Therefore, more studies should investigate the possibilities concerning how leaders can enhance resilience factors among their subordinates.

Furthermore, longitudinal studies would add much to our current knowledge about the long-term efficacy of resilience training programs. Most published studies focus on pre- and post-test. However, there is some evidence that broader effects of interventions may not be apparent immediately after the intervention.[166] In addition, posttest and short-term follow up measurements may be influenced by participants' increased awareness of their stress level and feelings due to increased self-reflection.

Furthermore, when learned knowledge and skills are not used, the training effect diminishes dramatically.[167] This reduction may be prevented by refresher courses. Already, such courses have been conducted as part of stress management training and long-term training effects have been found.[168] Furthermore, it would be useful to encourage military personnel to practice these skills while in military training, with the help and guidance of the supervisor to improve skills internalization.

Moderator Variables

A meta-analysis about the effectiveness of resilience training programs has shown that 50 % of the variance in resilience training effectiveness is due to moderators like training delivery format and design.[169] Future studies should examine additional moderators for training effectiveness, such as participants' attitudes, interests, and expectations.[170] Hence, one could examine the expectations of and satisfaction with the training itself as moderator variables.

Alternatively, one could examine the participants' motivation for resilience training and learning from this. Indeed, research in therapies has showed that motivation has a major influence on therapeutic efficacy and is central to behavioral change.[171] Because training is time-consuming and expensive in the military, one should examine the extent to which motivation influences the outcomes of resilience training, and how to improve motivation it.

Finally, many of the training modules require a self-reflective confrontation with oneself. Therefore, the influence of self-reflection on training effects should be considered in detail.

Outcome Variables

Some outcome variables, of great interest for military organizations, have been investigated scarcely, if at all. First, to our knowledge, the effect of resilience on biological variables has not been investigated. As mentioned above, stressors have a major impact on our body and can lead to many different physical and mental illnesses. The effects of psychological interventions at the physical level have already been shown in a study of behavioral cognitive stress management with civilians. Four months after the intervention, participants were administered a standardized stress test. Results showed that the intervention group had significant less stress hormones in the saliva than the control group.[172] Therefore, specific resilience-based interventions for military settings could be evaluated by measuring their effects on relevant biological variables.

Secondly, only a few studies have investigated the influence of individual differences on the outcome of resilience-based interventions. For example, individuals with a high self-esteem benefited more from resilience training than those with low self-esteem;[173] children at high-risk for depression showed greater resilience-building than those at low-risk;[174] and soldiers reporting higher level of exposure to mission stressors reported more improvement after a critical incident stress debriefing.[175] Determining who will benefit most from resilience interventions is a complex issue,[176] but such findings would be important in order to save money, time and resources.

Finally, to our knowledge, no study has investigated the influence of resilience building programs previously attended by military personnel while they are on active military duty or while undergoing standardized stress tests. Because resilience refers to the process of healthy functioning in the face of adversity and soldiers' resilience can be determined only when they are exposed to stress or trauma,[177] it is most important to measure the effects of resilience training in such situations.

Conclusion

The number and nature of stressors associated with military life increase the probability that military personnel will develop mental and physical health problems and that those affected will withdraw from military service.[178] This is a problem because it is a challenge to attract enough military personnel to fill the ranks.[179,180] Dropouts and withdrawals due to stress-related diseases should therefore be avoided, because in addition to the loss of personnel, they also entail considerable administrative effort and costs for the military organization.[181] Therefore, it would be advantageous for military organizations to identify the factors that can foster resilience in military settings. Indeed, training programs are a good alternative to decrease absenteeism and the number of military personnel who release.[182] Positive psychology offers the theoretical foundation for the promotion of positive emotions and individual strengths, so that as much as possible can be profited from the existing human potential. Ultimately, potential cost savings are much higher than training costs.[183]

Additionally, it is important to prevent recruits' attrition due to stress-related and trauma-related psychological and physical disorders. Prevention is not limited to offering resilience-based interventions to recruits but can also take place during selection. For example, it is possible to measure resilience and other resilience factors when applying to or entering a military organization. As such, military organizations could identify the recruits who would benefit most from resilience training and their respective training needs. In addition, the selection measurement could provide information on suitable occupations that would match applicants' resilience traits and stressors. Such matches when applied for some deployment positions result in a healthy warrior effect.[184] The effects of adapting work stressors to the individual could be doubled because adapting stressors to a person can be the first step to increased resilience. A long-term study showed that workplace challenge stressors predict an increase in resilience, whereas hindrance stressors predict a decline in resilience.[185]

Resilience is a central concept for the military community and is seen as an important component in keeping military personnel healthy and fit for service.[186] Therefore we should always keep in mind that: "*Waiting for illness or injury to occur is not the way commanders […] approach high-risk actions; and it is not the way we should approach high psychological risk activities.*"[187]

Notes

1 Peter Becker, *Gesundheit durch Bedürfnisbefriedigung* (2006). Göttingen: Hogrefe Verlag.
2 Susan Folkman et al., "Dynamics of a stressful encounter: Cognitive appraisal, coping, and encounter outcomes." *Journal of Personality and Social Psychology*, 50, no. 5 (1986): 992–1003, https://doi.org/10.1037/0022-3514.50.5.992

3 Richard S. Lazarus and Susan Folkman, *Stress, appraisal, and coping* (1984). New York: Springer Verlag.
4 Cited in Becker, " Gesundheit durch Bedürfnisbefriedigung".
5 Ibid.
6 Bruce S. McEwen, "Protection and damage from acute and chronic stress: Allostasis and allostatic overload and relevance to the pathophysiology of psychiatric disorders." *Annals of the New York Academy of Sciences*, 1032, no. 1 (2004), 1–7, https://doi.org/10.1196/annals.1314.001
7 Urs M. Nater and Ulrike Ehlert, "Stressabhängige körperliche Beschwerden." In Hans-Ulrich Wittchen and Jürgen Hoyer (Eds.), *Klinische Psychologie und Psychotherapie*, pp. 872–881 (2006). Heidelberg: Springer Verlag.
8 Jong-Sun Lee, Eun-Jeong Joo, and Kyeong-Sook Choi, "Perceived stress and self-esteem mediate the effects of work-related stress on depression." *Stress and Health*, 29, no. 1 (2013): 75–81, https://doi.org/10.1002/smi.2428
9 Charles W. Hoge et al., "Combat duty in Iraq and Afghanistan, mental health problems, and barriers to care." *New England Journal of Medicine*, 351, no. 1 (2004): 13–22, https://doi.org/10.1056/NEJMoa040603
10 James Griffith and Courtney West, "Master resilience training and its relationship to individual well-being and stress buffering among army national guard soldiers." *The Journal of Behavioral Health Services & Research*, 40, no. 2 (2013): 140–55, https://doi.org/10.1007/s11414-013-9320-8
11 Adam J. Vanhove et al., "Can resilience be developed at work? A meta-analytic review of resilience-building programme effectiveness." *Journal of Occupational and Organizational Psychology*, 89, no. 2 (2016): 278–307, https://doi.org/10.1111/joop.12123
12 Jared P. Reis et al., "Factors associated with discharge during marine corps basic training." *Military Medicine*, 172, no. 9 (2007): 936–941, https://doi.org/10.7205/MILMED.172.9.936
13 Raymond W. Novaco, Thomas M. Cook, and Irwin G. Sarason, "Military recruit training: An arena for stress-coping skills." In Donald Meichenbaum and Matt E. Jaremco (Eds.), *Stress reduction and prevention*, pp. 377–418 (1983). New York: Plenum Press.
14 Dennis McGurk et al., "Joining the ranks: The role of indoctrination in transforming civilians to service members." In Amy B. Adler, Carl A. Castro and Thomas W. Britt (Eds.), *Military life: The psychology of serving in peace and combat (Vol. 2): Operational stress*, pp. 13–31. Westport CT: Praeger Security International.
15 Novaco, Cook, and Sarason, *"Military recruit training,"* 377–418.
16 Hubert Annen and Simeon Frei, *"Stressors and coping strategies of recruits during basic training."* (Paper presented at the 53rd Annual Conference of the International Military Testing Association, IMTA, 2011).

17 Monique F. Crane et al., "Strengthening resilience in military officer cadets: A group-randomized controlled trial of coping and emotion regulatory self-reflection training." *Journal of consulting and clinical psychology*, 87, no. 2 (2019): 125–40, https://doi.org/10.1037/ccp0000356
18 Wim Kamphuis, Ward Venrooij, and Coen van den Berg. *A model of psychological resilience for the Netherlands Armed Forces* (2012). http://psyres mil.org/wp-content/uploads/2013/11/Kamphuis-et-al-A-model-of-psycho logical-resilience-for-the-NLD-AF-IMTA-2012.pdf
19 Hoge et al., *"Combat duty,"* 13–22.
20 Roy Eidelson, Mark Pilisuk, and Stephen Soldz, "The dark side of comprehensive soldier fitness." *American Psychologist*, 66, no. 7 (2011): 646–647, https://doi.org/10.1037/a0025272
21 Kamphuis, Venrooij, and Berg, *"A model of psychological resilience"*.
22 Avraham Bleich, Marc Gelkopf, and Zahava Solomon, "Exposure to terrorism, stress-related mental health symptoms, and coping behaviors among a nationally representative sample in Israel," *Jama*, 290, no. 5 (2003): 612–620, https://doi.org/10.1001/jama.290.5.612
23 Emmy E. Werner, Jessie M. Bierman, and Fern E. French. *The children of Kauai: A longitudinal study from the prenatal period to age ten* (1971). Honolulu: University of Hawaii Press.
24 Aaron Antonovsky, *Unraveling the mystery of health: How people manage stress and stay well* (1987). San Francisco, CA, US: Jossey-Bass.
25 Martin E.P. Seligman, "Positive psychology, positive prevention, and positive therapy." In Shane J. Lopez and Charles R. Snyder (Ed.), *Handbook of positive psychology*, pp. 3–12. New York: Oxford University Press.
26 Ibid.
27 Willibald Ruch and René T. Proyer, "Positive Interventionen: Stärkenorientierte Ansätze." In Renate Frank (Ed.), *Therapieziel Wohlbefinden: Ressourcen aktivieren in der Psychotherapie*, pp. 84–92. Heidelberg: Springer Verlag.
28 Nancy L. Sin and Sonja Lyubomirsky, "Enhancing well-being and alleviating depressive symptoms with positive psychology interventions: A practice-friendly meta-analysis." *Journal of Clinical Psychology*, 64, no.5 (2009): 467–487.
29 Martin E.P. Seligman and Mihaly Csikszentmihalyi, "Positive psychology: An introduction." *American Psychologist*, 55 (2000): 5–14, https://doi.org/10.1037/0003-066X.55.1.5
30 Sin and Lyubomirksy, *"Enhancing well-being,"* 467–487.
31 Michael D. Matthews, "Toward a positive military psychology." *Military Psychology*, 20 (2008): 289–298, https://doi.org/10.1080/08995600802345246
32 Nadine Eggimann and Hubert Annen, "Positive Psychologie im Militär." *Allgemeine Schweizerische Militärzeitschrift*, 03 (2020): 35–37.
33 Ann S. Masten, *Ordinary magic: Resilience in development* (2014). New York: The Guilford Press.

34 Andrea Chmitorz et al., "Intervention studies to foster resilience – A systematic review and proposal for a resilience framework in future intervention studies." *Clinical psychology Review*, 59 (2018): 78–100, https://doi.org/10.1016/j.cpr.2017.11.002

35 Kathryn M. Connor and Jonathan R. Davidson, "Development of a new resilience scale: The Connor Davidson Resilience Scale (CD-RISC)." *Depression and Anxiety*, 18, no. 2 (2003): 76–82, https://doi.org/10.1002/da.10113

36 Raffael Kalisch et al., "The resilience framework as a strategy to combat stress-related disorders." *Nature Human Behavior*, 1, no. 11 (2017): 784–790, https://doi.org/10.1038/s41562-017-0200-8

37 Masten, *"Ordinary magic: Resilience in development"*.

38 "The road to resilience", American Psychological Association, accessed July 20, 2019, http://www.apa.org/helpcenter/road-resilience.aspx

39 Gill Windle, Kate M. Bennett, and Jane Noyes, "A methodological review of resilience measurement scales." *Health and Quality of Life Outcomes*, 9, no. 1 (2011): 8, https://doi.org/10.1186/1477-7525-9-8

40 George A. Bonanno, "Loss, Trauma, and Human Resilience: Have we underestimated the human capacity to thrive after extremely adverse events?," *American Psychologist*, 59, no. 1 (2004): 20–28, https://doi.org/10.1037/0003-066X.59.1.20

41 Vanhove et al., *"Can resilience be developed at work?,"* 278–307.

42 Steven M. Southwick et al., "Resilience definitions, theory, and challenges: Interdisciplinary perspectives," *European Journal of Psychotraumatology*, 5, no. 1 (2014), https://doi.org/10.3402/ejpt.v5.25338

43 Ann S. Masten, "Ordinary magic: Resilience processes in development," *American Psychologist*, 56, no. 3 (2001): 227–238, https://doi.org/10.1037//0003-066X.56.3.227

44 Kerry A. Sudom, Jennifer E. C. Lee, and Mark A. Zamorski, "A longitudinal pilot study of resilience in Canadian military personnel." *Stress and Health Journal of the International Society for the Investigation of Stress*, 30, no. 5 (2014): 377–85, https://doi.org/10.1002/smi.2614

45 Masten, *"Ordinary magic,"* 921–930.

46 Chmitorz et al., *"Intervention studies to foster resilience,"* 78–100.

47 Kalisch et al., *"The resilience framework,"* 784–790.

48 Ann S. Masten, "Resilience in developing systems: Progress and promise as the fourth wave rises." *Development and Psychopathology*, 19, no. 3 (2007): 921–930, https://doi.org/10.1017/S0954579407000442

49 Tracy O. Afifi and Harriet L. MacMillan, "Resilience following child maltreatment: A review of protective factors." *The Canadian Journal of Psychiatry*, 56, no. 5 (2011): 266–272, https://doi.org/10.1177/070674371105600505

50 Lü, Wei et al., "Resilience as a mediator between extraversion, neuroticism and happiness, PA and NA." *Personality and Individual Differences*, 63 (2014): 128–33, https://doi.org/10.1016/j.paid.2014.01.015
51 Affifi and MacMillan, *"Resilience following child maltreatment,"* 266–272.
52 Lee, Joo, and Choi, *"Perceived stress and self-esteem,"* 75–81.
53 Özlem Karairmak, *"Investigation of Personal Qualities Contributing to Psychological Resilience among Earthquake Survivors: A Model Testing Study."* (Unpublished PhD thesis, Ankara: Middle East Technical University, 2007).
54 David Fletcher and Mustafa Sarkar, "Psychological resilience: A review and critique of definitions, concepts, and theory." *European Psychologist*, 18, no. 1 (2013): 12–23, https://doi.org/10.1027/1016-9040/a000124
55 Thomas J. Dishion and Arin Connell, "Adolescents' resilience as a self-regulatory process: Promising themes for linking intervention with developmental science." In Barry M. Lester, Ann Masten, and Bruce McEwen (Ed.), *Resilience in children*, pp. 125–138 (2006). Boston: New York Academy of Sciences.
56 Lisa S. Meredith et al., "Promoting psychological resilience in the U.S. military." *Rand Health Q*, 1, no. 2 (2011), https://www.rand.org/con-tent/dam/rand/pubs/monographs/2011/RAND_MG996.pdf.
57 Annen, *"Resilienz,"* 24–35.
58 Meredith et al., *"Promoting psychological resilience"*.
59 Affifi and MacMillan, *"Resilience following child maltreatment,"* 266–272.
60 Meredith et al., *"Promoting psychological resilience"*.
61 Hubert Annen, "Resilienz – eine Bestandsaufnahme." *Military Power Revue der Schweizer Armee*, 1/2017 (2017): 24–35.
62 Windle et al., *"A methodological review,"* 8.
63 Chmitorz et al., *"Intervention studies to foster resilience,"* 78–100.
64 Windle et al., *"A methodological review,"* 8.
65 Bruce W. Smith et al., "The brief resilience scale: Assessing the ability to bounce back." *International Journal of Behavioral Medicine*, 15, no. 3 (2008): 194–200, https://doi.org/10.1080/10705500802222972
66 Connor and Davidson, *"Development of a new resilience scale,"* 76–82.
67 Oddgeir Friborg et al., "A new rating scale for adult resilience: What are the central protective resources behind healthy adjustment?" *International Journal of Methods in Psychiatric Research*, 12, no. 2 (2003): 65–76, https://doi.org/10.1002/mpr.143
68 Lynda A. King and Daniel W., "Measuring resilience and growth." In Brett A. Moore and Jeffrey E. Barnett (Ed.), *Military psychologists' desk reference*, pp. 302–305 (2013). New York: Oxford University Press.
69 Paul T. Bartone *"A short hardiness scale."* (Paper presented at the Annual Convention of American Psychological Society, New York, 1995), https://apps.dtic.mil/dtic/tr/fulltext/u2/a298548.pdf.

70 Douglas C. Johnson et al., "Development and initial validation of the response to stressful experiences scale." *Military Medicine*, 176, no. 2 (2011): 161–169, https://doi.org/10.7205/MILMED-D-10-00258
71 Meredith et al., *"Promoting psychological resilience"*.
72 Crane et al., *"Strengthening resilience,"* 125–140.
73 Meredith et al., *"Promoting psychological resilience"*.
74 Madlaina Niederhauser, Caroline Huber, and Hubert Annen, "Der Einfluss von Resilienz auf die militärische Leistung." *Allgemeine Schweizerische Militärzeitschrift*, 3/2016, (2016): 48–49, https://doi.org/10.5169/seals-587026
75 Lew Hardy et al., "The relationship between transformational leadership behaviors, psychological, and training outcomes in elite military recruits." *The Leadership Quarterly*, 21, no. 1 (2010): 20–32, https://doi.org/10.1016/j.leaqua.2009.10.002
76 Hardy et al., *"The relationship between transformational leadership,"* 20–32.
77 Paul B. Lester et al., "Evaluation of relationships between reported resilience and soldier outcomes – Report no. 2: Positive performance outcomes in officers (promotions, selections, & professions)." (2011), http://digitalcommons.unl.edu/publicpolicypublications/129
78 Madlaina Niederhauser, Regula Züger, and Hubert Annen, "Ein Resilienztraining für die Schweizer Armee auf dem Prüfstand." *Allgemeine Schweizerische Militärzeitschrift*, 10/2017 (2017): 40–43.
79 Thomas Wyss and Hubert Annen, *"Studie progress."* Report, Magglingen/Birmensdorf, (2013).
80 Kamphuis, Venrooij, and van der Berg, *"A model of psychological resilience"*.
81 Karen J. Reivich, Martin E. P. Seligman, and Sharon McBride, "Master resilience training in the U.S. Army." *The American Psychologist*, 66, no. 1 (2011): 25–34, https://doi.org/10.1037/a0021897
82 Kamphuis, Venrooij, and van der Berg, *"A model of psychological resilience"*.
83 Bonnano, *"Loss, Trauma, and Human Resilience,"* 20–28.
84 Sudom, Lee, and Zamorski, *"A longitudinal pilot study of resilience,"* 377–385.
85 Vanhove et al., *"Can resilience be developed at work?,"* 278–307.
86 Steven M. Brunwasser, Jane E. Gillham, and Eric S. Kim, "A meta-analytic review of the Penn Resiliency Program's effect on depressive symptoms." *Journal of Consulting and Clinical Psychology*, 77, no. 6 (2009): 1042–54, https://doi.org/10.1037/a0017671
87 Vanhove et al., *"Can resilience be developed at work?,"* 278–307.
88 Lazarus and Folkman, "Stress, appraisal, and coping".
89 Aaron T. Beck. *Cognitive therapy and the emotional disorders* (1976). New York: International University Press.
90 Albert Ellis, *Reason and emotion in psychotherapy* (1962). Oxford, England: Lyle Stuart.

91 Martin E. P Seligman, "Building Resilience." *Harvard Business Review*, 89, no. 4 (2011): 100–106.
92 Vanhove et al., *"Can resilience be developed at work?,"* 278–307.
93 Reivich, Seligman, and McBride, *"Master resilience training,"* 25–34.
94 Brunwasser, Gillham, and Kim, *"A meta-analytic review,"* 1042–1054.
95 Vanhove et al., *"Can resilience be developed at work?,"* 278–307.
96 Niederhauser, Züger, and Annen, *"Ein Resilienztraining für die Schweizer Armee,"* 40–43.
97 Reivich, Seligman, and McBride, *"Master resilience training,"* 25–34.
98 Brunwasser, Gillham, and Kim, *"A meta-analytic review,"* 1042–1054.
99 Martin E. P. Seligman, Randal M. Ernst, Jane Gillham, Karen Reivich, and Mark Linkins, "Positive education: Positive psychology and classroom interventions." *Oxford Review of Education*, 35, no. 3 (2009): 293–311, https://doi.org/10.1080/03054980902934563
100 Vanhove et al., *"Can resilience be developed at work?,"* 278–307.
101 Chmitorz et al., *"Intervention studies to foster resilience,"* 78–100.
102 Frank W. Bond and David Bunce, "Mediators of change in emotion-focused and problem-focused worksite stress management interventions." *Journal of Occupational Health Psychology*, 5, no. 1 (2000): 156–63, https://doi.org/10.1037//1076-8998.5.1.156
103 Fred Luthans et al., "The development and resulting performance impact of positive psychological capital." *Human Resource Development Quarterly*, 21, no. 1 (2010): 41–67, https://doi.org/10.1002/hrdq.20034
104 Anthony M. Grant, Linley Curtayne, and Geraldine Burton, "Executive coaching enhances goal attainment, resilience and workplace well-being: A randomized controlled study." *The Journal of Positive Psychology*, 4, no. 5 (2009): 396–407, https://doi.org/10.1037/a0019212
105 Paul R. Grime, "Computerized cognitive behavioral therapy at work: A randomized controlled trial in employees with recent stress-related absenteeism." *Occupational medicine (Oxford, England)*, 54, no. 5 (2004): 353–59, https://doi.org/10.1093/occmed/kqh077
106 Salvatore R. Maddi, Stephen Kahn, and Karen L. Maddi, "The effectiveness of hardiness training." *Consulting Psychology Journal: Practice and Research*, 50, no. 2 (1998): 78–86, https://doi.org/10.1037/1061-4087.50.2.78
107 Grant, Curtayne, and Burton, *"Executive coaching,"* 396–407.
108 Nicola W. Burton, Ken I. Pakenham, and Wendy J. Brown, "Feasibility and effectiveness of psychosocial resilience training: A pilot study of the READY program." *Psychology, Health & Medicine*, 15, no. 3 (2010): 266–277, https://doi.org/10.1080/13548501003758710
109 Grant, Curtayne, and Burton, *"Executive coaching,"* 396–407.
110 Grime, *"Computerized cognitive behavioral therapy at work,"* 353–359.

111 Burton Pakenham, and Brown, *"Feasibility and effectiveness of psychosocial resilience training,"* 266–277.
112 Luthans et al., *"The development and resulting performance impact,"* 41–67.
113 Bond and Bunce, *"Mediators of chance,"* 156–163.
114 Burton, Pakenham, and Brown, *"Feasibility and effectiveness of psychological resilience training,"* 266–277.
115 Jo-Anne Abbott et al., "The impact of online resilience training for sales managers on well-being and performance." *Electronic Journal of Applied Psychology*, 5 (2009): 89–95, https://pdfs.semanticscholar.org/923 f/6aeaededcc55d8e0143d3b80bf6376f03ae1.pdf
116 Timothy D. Hodges, *"An Experimental Study of the Impact of Psychological Capital on Performance, Engagement, and the Contagion Effect."* (PhD thesis, University of Nebraska, 2010), http://digitalcommons.unl.edu/businessdiss/7
117 Bengt B. Arnetz et al., "Trauma resilience training for police: Psychophysiological and performance effects." *Journal of Police and Criminal Psychology*, 24, no. 1 (2009): 1–9, https://doi.org/10.1007/s11896-008-9030-y
118 Rollin McCraty and Mike Atkinson, "Resilience training program reduces physiological and psychological stress in police officers." *Global Advances in Health and Medicine*, 1, no. 5 (2012): 44–66, https://doi.org/10.7453/gahmj.2012.1.5.013
119 Arnetz et al., *"Trauma resilience training,"* 1–9.
120 Hodges, *"A experimental study of the impact of psychological capital"*.
121 Ibid.
122 Arnetz et al., *"Trauma resilience training,"* 1–9.
123 R.D. Rose, J.C. Buckey, T.D. Zbozinek, S.J. Motivala, D.E. Glenn, J.A. Cartreine, and M.G. Craske, A randomized controlled trial of a self-guided, multimedia, stress management and resilience training program. *Behaviour Research and Therapy*, 51, no. 2 (2013): 106–112, https://doi.org/10.1016/j.brat.2012.11.003
124 Jane Gillham et al., *The Penn Resiliency Program* (Unpublished Manual, University of Pennsylvania, Philadelphia, 1990).
125 Seligman et al., *"Positive education,"* 293–311.
126 Brunwasser, Gillham, and Kim, *"A meta-analytic review,"* 1042–1054.
127 Meredith et al., *"Promoting psychological resilience"*.
128 Amy B. Adler et al., "Battlemind debriefing and battlemind training as early interventions with soldiers returning from iraq: Randomization by platoon." *Journal of Consulting and Clinical Psychology*, 77, no. 5 (2009): 928–40, https://doi.org/10.1037/a0016877
129 For an overview see Kathleen Mulligan et al., "Psycho-educational interventions designed to prevent deployment-related psychological

ill-health in Armed Forces personnel: A review." *Psychological Medicine*, 41, no. 4 (2011): 673–686, https://doi.org/10.1017/s003329171000125x
130 Adler et al., *"Battlemind debriefing,"* 928–940.
131 Carl Andrew Castro, Amy B. Adler, Dennis McGurk, and Paul D. Bliese, "Mental health training with soldiers four months after returning from Iraq: Randomization by platoon." *Journal of Traumatic Stress*, 25, no. 4 (2012): 376–383, https://doi.org/10.1002/jts.21721
132 Adler et al., *"Battlemind debriefing,"* 928–940.
133 Brett T. Litz et al., "A randomized, controlled proof-of-concept trial of an internet-based, therapist-assisted self-management treatment for posttraumatic stress disorder." *American Journal of Psychiatry*, 164, no. 11 (2007): 1676–1683, https://doi.org/10.1176/appi.ajp.2007.06122057
134 Martha Kent et al., "A resilience-oriented treatment for posttraumatic stress disorder: Results of a preliminary randomized clinical trial." *Journal of Traumatic Stress*, 24, no. 5 (2011): 591–595, https://doi.org/10.1002/jts.20685
135 David J. Linkh and Scott M. Sonnek, "An application of cognitive-behavioral anger management training in a military/occupational setting: Efficacy and demographic factors." *Military Medicine*, 168, no. 6 (2003): 475–478, https://doi.org/10.1093/milmed/168.6.475
136 Jeffrey A. Cigrang, Sandy L. Todd, and Eric G. Carbone, "Stress management training for military trainees returned to duty after a mental health evaluation: Effect on graduation rates." *Journal of Occupational Health Psychology*, 5, no. 1 (2000): 48–55, https://doi.org/10.1037/1076-8998.5.1.48
137 Arthur Williams et al., "STARS: Strategies to assist navy recruits' success." *Military Medicine*, 172, no. 9 (2007): 942–949, https://doi.org/10.7205/MILMED.172.9.942
138 Arthur Williams et al., "Psychosocial effects of the BOOT STRAP intervention in Navy recruits." *Military Medicine*, 169, no. 10 (2004): 814–820, https://doi.org/10.7205/MILMED.169.10.814
139 Andrew Cohn and Ken Pakenham, "Efficacy of a cognitive-behavioral program to improve psychological adjustment among soldiers in recruit training." *Military Medicine*, 173, no. 12 (2008): 1151–1157, https://doi.org/10.7205/MILMED.173.12.115.1
140 Amy B. Adler et al., "Resilience training with soldiers during basic combat training: Randomisation by platoon." *Applied Psychology: Health and Well-Being*, 7, no. 1 (2015): 85–107, https://doi.org/10.1111/aphw.12040
141 Norman Jones et al., "Resilience-based intervention for UK military recruits: A randomised controlled trial." *Occupational and Environmental Medicine*, 76, no. 2 (2019): 90–96, https://doi.org/10.1136/oemed-2018-105503
142 Crane et al., *"Strengthening resilience,"* 125–140.

143 Niederhauser, Züger, Annen, "Ein Resilienztraining für die Schweizer Armee," 40–43.
144 Chantal Utzinger. *Einfluss eines Resilienztrainings auf akuten Stress und Stressreaktivität* (Unpublished Master thesis, Zuerich: University of Zuerich, 2017).
145 Roman W. Spinnler. *Army Resilience Training bei der SWISSCOY. Erkenntnisse aus einer Feldstudie zum Resilienztraining bei SWISSINT für einen Friedensförderungseinsatz der Schweizer Armee im Kosovo* (Unpublished Master thesis, Bern: University Bern, 2019).
146 Griffith and West, *"Master resilience training,"* 140–155.
147 Cornum, Matthews, and Seligman, *"Comprehensive soldier fitness".*
148 Ibid.
149 George W. Casey, "Comprehensive soldier fitness: A vision for psychological resilience in the U.S. Army." *The American Psychologist*, 66, no. 1 (2011): 1–3, https://doi.org/10.1037/a0021930
150 Ibid.
151 Seligman, *"Building resilience,"* 100–106.
152 Eidelson, Pilisuk, and Soldz, *"The dark side of comprehensive soldier fitness,"* 646–647.
153 Griffith and West, *"Master resilience training and its relationship to individual well-being and stress buffering among army national guard soldiers,"* 140–155.
154 Peter D. Harms et al., *"Report #4: Evaluation of Resilience Training and Mental and Behavioral Health Outcomes Peter. The Comprehensive Soldier Fitness Program Evaluation."* (2013), http://digitalcommons.unl.edu/cgi/viewcontent.cgi?article=1009&context=pdharms.
155 Paul B. Lester et al., "The comprehensive soldier fitness program evaluation. *Report # 3: Longitudinal Analysis of the Impact of Master Resilience Training on Self-Reported Resilience And Psychological Health Data,"* (2011), http://www.dtic.mil.
156 Vanhove et al., *"Can resilience be developed at work?,"* 278–307.
157 Ibid.
158 ibid.
159 Martin Fishbein, "Great expectations, or do we ask too much from community-level interventions?." *American Journal of Public Health*, 86, no. 8 (1996): 1075–1076, https://doi.org/10.2105/AJPH.86.8_Pt_1.1075
160 Vanhove et al., *"Can resilience be developed at work?,"* 278–307.
161 Thomas W. Britt et al., "Effective and ineffective coping strategies in a low-autonomy work environment." *Journal of Occupational Health Psychology*, 21, no. 2 (2016): 154–68, https://doi.org/10.1037/a0039898
162 Vanhove et al., *"Can resilience be developed at work?,"* 278–307.
163 Crane et al., *"Strengthening resilience in military officer cadets,"* 125–140.
164 Lester et al., *"The comprehensive soldier fitness program evaluation. Report # 3".*

165 Hoge et al., "*Combat duty in Iraq and Afghanistan*," 13–22.
166 Abbott et al., "*The impact of online resilience training for sales managers*," 89–95.
167 Winfred Arthur Jr et al., "Factors that influence skill decay and retention: A quantitative review and analysis." *Human Performance*, 11, no. 1 (1998): 57–101, https://doi.org/10.1207/s15327043hup1101_3
168 Heribert Limm et al., "Stress management interventions in the workplace improve stress reactivity: A randomised controlled trial." *Occupational and Environmental Medicine*, 68, no. 2 (2010): 126–133, https://doi.org/10.1136/oem.2009.054148
169 Vanhove et al., "*Can resilience be developed at work?*," 278–307.
170 Raymond A. Noe and Neil Schmitt, "The influence of trainee attitudes on training effectiveness: Test of a model." *Personnel Psychology*, 39, no. 3 (1986): 497–523, https://doi.org/10.1111/j.1744-6570.1986.tb00950.x
171 Johan Y. Y. Ng et al., "Self-determination theory applied to health contexts: A meta-analysis." *Perspectives on Psychological Science*, 7, no. 4 (2012): 325–340, https://doi.org/10.1177/1745691612447309
172 Karin Hammerfald et al., "Persistent effects of cognitive-behavioral stress management on cortisol responses to acute stress in healthy subjects – A randomized controlled trial." *Psychoneuroendocrinology*, 31, no. 3 (2006): 333–339, https://doi.org/10.1016/j.psyneuen.2005.08.007
173 Samuel Felder, "*Der Einfluss des Selbstwertgefühls auf die Entwicklung verschiedener Lernparameter in einem Resilienztraining der Schweizer Armee*," (Unpublished Master thesis, Bern: University of Berne, 2018).
174 Brunwasser, Gillham, and Kim, "*A meta-analytic review of the Penn Resiliency Program's effect on depressive symptoms*," 1042–1054.
175 Adler et al., "*Battlemind debriefing and battlemind training as early interventions with soldiers returning from Iraq: Randomization by platoon*," 928–940.
176 Vanhove et al., "*Can resilience be developed at work?*," 278–307.
177 Chmitorz et al., "*Intervention studies to foster resilience*,", 78–100.
178 Reis et al., "*Factors associated with discharge during Marine Corps basic training*," 936–941.
179 Niederhauser, Züger, and Annen, "*Ein Resilienztraining für die Schweizer Armee*," 40–43.
180 Beth J. Aesch, *Navigating current and emerging army recruiting challenges: What research tell us?*. Santa Monica: RAND Corporation.
181 Michael D. Matthews, Richard M. Lerner and Hubert Annen, " Noncognitive amplifiers of human performance: Unpacking the 25/75 rule." In Michael D. Matthews and David M. Schnyder (Eds.), *Human performance optimization*, pp. 356–382 (2019). Oxford: University Press.
182 Sudom, Lee, and Zamorski, "*A longitudinal pilot study of resilience*," 377–385.
183 Williams et al., "*STARS: Strategies to assist navy recruits' success*," 942–949.

184 Gerald E. Larson, Robyn M. Highfill-McRoy, and Stephanie Booth-Kewley, "Psychiatric diagnoses in historic and contemporary military cohorts: Combat deployment and the healthy warrior effect." *American Journal of Epidemiology*, 167, no. 11 (2008): 1269–1276, https://doi.org/10.1093/aje/kwn084
185 Monique F. Crane and Ben J. Searle, "Building resilience through exposure to stressors: The effects of challenges versus hindrances." *Journal of Occupational Health Psychology*, 21, no. 4 (2016): 468–479, https://doi.org/10.1037/a0040064
186 Meredith et al., *"Promoting psychological resilience"*.
187 Reivich, Seligman, and McBride, *"Master resilience training in the U.S.Army,"* 25–34.

Liisi Toom

Built-in Resilience Intervention in Academic Military Education

If the COVID-19 outbreak has demonstrated anything, it is that ambiguous, stressful, unexpected, and potentially lethal events can happen in an instant and hardly anyone is prepared. McGraw et al.[1] say that a proactive approach to psychological health promotion practices among soldiers should be preferred to psychological sequelae[2] treatment plans. In 2020, drawing from the wish to build resilience in cadets and using the concepts of sports and positive psychology, the author of this article designed a resilience intervention for the "Applications of Psychology in the Defence Forces" course. Although the course tackles mental preparedness issues, the COVID-19 outbreak demanded a more proactive action for which the course was a suitable medium. The aim of the intervention was to help cadets cope with the stress caused by the outbreak, build stronger awareness of the mental health discourse, and give practical advice for building resilience in these adverse times. To date, two groups of students have undergone this pilot intervention (in spring of 2020 and January–February of 2021). Today, the objective of this unique mental health intervention in the Estonian Military Academy is to pave the way for a more integrated mental resilience training program for military education. This article provides recommendations to military institutions that are considering integrating psychology for mental well-being and general performance enhancement and gives examples on how mental health and resilience interventions could be used more broadly in everyday military life. It is also the aim of the author to encourage military leaders, as well as educators and cadets, to be open to integrating psychology more openly and to notice real-life opportunities to provide naturally occurring stress inoculation possibilities with healthy tools to encourage effective coping and grow mental well-being.

Mental Strength Integrated

It is commonly known that joining a new team or organization is stressful and challenging. People are unfamiliar, our everyday rhythms may change, and we are faced with a new culture along with new professional challenges. Good onboarding[3] and mentorship are vital to making a smooth transition so that a new person can start using his/her professional skills and perform. For the military, indoctrination is a well thought out process to help individuals become part of the unified, strong, and seamlessly performing team. Excess stress from changes in identity and new experience can accumulate. Stress may also come from one's personal life or culture, plus the behaviour, and attitudes of a new team, making it difficult to cope. If a person lacks coping mechanisms, this added stress may affect their well-being. Performance depends on the ability to manage change and direct one's focus on a task at hand rather than emotions. While seasoned professionals may be able to adapt with changes more easily, junior level professionals and new recruits need direction and an open environment. This is also important under circumstances such as the COVID-19 outbreak where stress is continuous and overwhelming, affecting every aspect of a person's life. Even resilient people need more care and a supportive environment where they can feel safe and in control. This is especially applicable in military environments where excess stress, challenges, and ever-changing tasks are common.

In stressful environments, including basic training, and changing circumstances, an intervention for new cadets to establish or strengthen mental patterns and coping mechanisms is highly recommended. A program dedicated to understanding stress and building relationships, awareness, and practical coping mechanisms is vital in any context. Having such a program should be adjustable with any onboarding program for employees and military cadets at any level, giving them a warm welcome and creating a common foundation for stronger mental performance. This would provide cadets who may not have the strongest self-awareness or have unresolved issues that may become visible only in high-stress situations with communal support and acceptance, to rise to the occasion. A person must first have the professional skills, but these are rarely present in new cadets, and this creates additional stress. Learning about stress mechanisms and what to do to ensure a sense of well-being is important for recruits as well as civilian employees/leaders. This is especially the case where motivations for recruitment are clearly external. It is not uncommon that a desire for stable income is the motivation, or military service may seem to be the only option for cadets. Developing these coping skills is also critical when a person lacks strong internal motivation to do what military personnel are expected to do: act decisively in peace as well as war. Research by Hourani *et al* has shown that stress causes more performance loss in junior positions and lower paygrades[4]. Though high-end leaders and junior cadets work on different levels, the concept of stress is universal, making mental preparations an inevitable requirement. According to Hourani and colleagues, 32.2 % of respondents reported that stress is highest at work and 18.7 % said that stress is highest within the family. Furthermore, 28 % reported that these stressors interfered with their ability to carry out the requirements of their military position "a lot"[5].

The same study revealed that the youngest, and lowest ranking, personnel experience higher levels of stress, suffer more mental health issues, and are less productive compared to older and higher-ranking personnel. Findings like these show that there is a need to develop coping strategies and stress management techniques early in a military career[6]. A preventive approach proved to be effective in a study conducted among navy recruits; a cognitive-behavioral intervention improved functioning and training performance, and reduced attrition among recruits with higher perceived levels of stress and depression[7]. This demonstrates that interventions, programs, and "communication" that help to manage stress and build resilience in the military could be a performance enhancement tool that is equal to, or even more important, than technical training. According to the toolbox for young cadets, "Fundamentals of warfighting for ground forces of Estonian Defence Forces," no technical tools or equipment can provide superiority if a person is not motivated to use them[8]. We may gather from these excerpts that *motivation to act* is not a talent – nobody is born with it – and excessive stress that one cannot cope with can eliminate this motivation even from the most heroic person.

This author believes that stress is a normal part of a military environment and members are expected to live, perform, and get results with growing demands on their mental and technical skills. This applies whether they are recruits or top military leaders, and learning to integrate mental techniques, as well as managing personal and team well-being must be integrated into daily life, education, and communication. As time and money are often scarce, integration through understanding what builds well-being and resilience and practicing it from personal life to military specific training and leadership styles is the easiest and most accessible way to provide extensive results in well-being of the personnel in the organization.

COVID-19 and Educational Environment

Military operations involve five common stress factors. We can categorize them as follows: (1) isolation; (2) uncertainty; (3) inability to change a situation; (4) boredom, routine; and (5) sense of fear and danger. It is worth noting that the stressors related to COVID-19 coincide with the stressors of a warzone. Cipriano et al.[9] describe the emotional stressors in a brief about COVID-19 as follows: (1) health-related fear, general fear that a person or their family member may contract COVID-19 or have their physical health jeopardized; (2) anxiety about day-to-day management, including supporting him- or herself and the family while working full-time from home and adapting with new technologies for teaching and learning; (3) leaders share an additional stress of knowing the gravity of their decisions and impact on the school community; and (4) students report more anxiety and concern about future education, college, and career plans. Similar concerns were evident in the Estonian Military Academy during the first wave of COVID-19. All those involved adjusted quickly with e-learning and re-adjusted with working from home and following the safety regulations. Plus, there was clear communication about how the Academy would support isolation or what to do when health-related issues (COVID) present themselves. However, there was little to no information in the Academy about supporting emotional well-being and mental health. It is a common presumption that military personnel are mentally and physically strong and, thus, ready for crisis situations. Even if this is true, extended mental support is necessary with additional and long-lasting stressors. The mental well-being brief by Cipriano et al. also stated that self-awareness, including the ability to recognize and identify our emotions, is the first step toward understanding how emotions can influence our thinking, decisions, and behavior[10]. This proposition underlined the activities of the current intervention for creating well-being in stressful times, overcoming environmental challenges, and providing the tools for similar circumstances in the future.

Cipriano *et al.*[11] explain that dysregulated emotions can inhibit healthy relationships. For example, these unhealthy relationships can develop among teachers and their students or between family members. This also applies for leaders and subordinates. If we do not manage our emotions effectively, we will be unable to effectively teach, learn, parent, or lead. Thriving through a pandemic, or similar unexpected challenges on a smaller or wider scale, requires a healthy mental flexibility that is best achieved by experiencing more pleasant than unpleasant emotions, and enabling resilience during and after traumatic events. This is a basic notion that teachers and leaders of the Academy can develop and pass on to cadets to help them enhance their mental strength. This is also something that can be an integral part of a culture that appreciates and values people and is necessary for members to thrive in highly stressful situations. While military culture is about strength, this strength can be enhanced by people being able to understand and regulate emotions and support each other.

Stress, Well-being, and Performance in the Military

In recent years, and with the increasing popularity of positive psychology practices, a connection has been made between performance, stress, and well-being. Today, we can follow West-Point Academy's mental training practices on Instagram, and see that mental skills are an asset and a requirement for higher performance. Throughout the organization, valuing personnel is widely acknowledged, yet the best practices and techniques for supporting their well-being are still just being developed. For these initiatives to have greater impact, leaders must ensure that the practices are well integrated into daily life, attitudes, and educational programs. Therefore, in a military organization, as well as in other high-performance cultures, it is important to learn how to sustain well-being both physically (sleep, nutrition, supercompensation) and mentally (emotional regulation, intra- and inter-relationships). According to Williams et al.[12], maladaptive responses to stress interfere most with a recruit's ability to initiate or sustain interpersonal relationships, master a required content and skills, manage feelings and moods, and cognitively control the interpretation of information and experience. Evidently, body and mind need to coincide and work in harmony.

Yet, various sources and discussions with military leaders and educators have shown that the use of mental skills from breathing to self-motivation is increasingly topical but less emphasis is put into their execution. Mental skills require practicing just like a physical workout. This practice takes time and should be thus valued enough by curriculum creators and leaders who decide what a soldier, cadet or a recruit should do with their daily schedules. We do see this happening in Estonia, where psychology courses were added 5 ECTSs (European Credit Transfer and Accumulation System) in total as of 2021 (previously 7 EAP in total, now 12 EAP).

It is important to note that high levels of stress are not pervasive only to the military personnel who are deployed or exposed to combat. Many cadets may never become suitable for deployment because their fundamental mental training is lacking, and they drop out for not being "mentally strong enough." Hourani et all found that with military downsizing there are increased duties, high levels of stress may be pervasive among all military personnel.[13] Since the world is getting more complicated, and challenges are becoming more ambiguous, it is expected that general stress levels will rise. While soldiers preparing for deployment usually participate in some stress management programs, "regular" personnel are expected to figure this out by themselves.

The author of this article suggests that in contemporary working environments, responsibility for this sort of development should be transferred from an individual to an organization.

Preparation for Stress and Popular Resilience Programs

Martínez-Sánchez[14] notes that contingent personnel are already psychologically prepared and trained during selection and assessment to reinforce their capacity to cope with the problems and difficulties that they will encounter during operations. Advanced resilience training is most often applied only before a mission. Martínez-Sánchez also stresses how important it is to give additional psychological support to the military personnel by counselling the commanders on the functions and applicability of military psychology, informing the chiefs and officers of contingent units about aspects such as motivation and the suitability of people for the tasks assigned to them, the morale of the unit, prevention of psychosocial risks, improvement of working conditions, planning leisure and free time, and guidelines and measures for detecting and preventing maladaptive behaviors. Even if no other intervention or program is applied, the three basic programs (selecting the fittest, preparing for challenges through physical training and counselling, and teaching necessary skills to the leaders) are still present. Though this provides extensive understanding on psychological effects, it could be deepened with real experience and training solely focused on psychological skills.

The most popular civilian resilience programs were developed in the University of Pennsylvania; they were also used for creating BOOT STRAP. The Comprehensive Soldier Fitness Program is a longitudinal project that has confirmed the efficiency of mental training for resilience over a long period[15]. Master Resilience Training is an intervention within the Comprehensive Soldier Fitness Program and a cornerstone for military resilience and a psychological health development initiative created in cooperation with the University of Pennsylvania; it includes 80 hours of classroom training for six core competencies: self-awareness, self-regulation, optimism, mental agility, strength of character, and connection. The program sees resilience as something one can learn so that a person can continue functioning normally even in situations of crisis.

While each program defines and lists the competencies and skills that suit a relevant culture within a team or an organization, all are rooted in performance/sports psychology and positive psychology research outcomes. One of the key terms used for training and developing resilience and mental well-being is *growth mindset*[16] that helps to see situations as challenges instead of problems. Leaders with a growth mindset see cadets as people capable of learning and developing and know how to properly support them. Another concept used in building resilience in military cohorts is *grit*,[17] a combination of perseverance and passion, helping to overcome difficult situations and obtain a variety of skills that help a person to become a specialist and a professional in their individual field over a long period of time. *Deliberate training* is a term used in sports psychology, stating that to achieve the best performance, every moment of training must be meaningful. These all are part of *psychological skills training* that provides special techniques for combining a physical performance with mental capacities. Psychological skills refer to setting a goal, motivation, identity, managing emotions, self-talk, visualization, and other capacities that direct our movements and behavior and can be trained to improve technical skills, speed, precision, balance, etc.

Resilience Programs at the Estonian Military Academy and in the Defence Forces

Currently, there are no distinct resilience programs used at the Estonian Military Academy nor in other structures under the Ministry of Defence. Stress-related topics are touched upon in preparation for soldiers who go on missions and, to some extent, during preparations for medical personnel. Although psychological aspects are mentioned in various courses in the curricula of the Estonian Defence Academy, topics like well-being, positive attitude, and resilience depend on the individual knowledge of soldiers and the skills of leaders. The organization and all the personnel involved with psychology are introduced to the Total Force Fitness concept[18] as a guideline. Unfortunately, awareness and daily appliance of these guidelines is inconsistent throughout the organization. Considering that military recruits are healthy youngsters placed in stressful, demanding, and unfamiliar situations, well-being should be central to avoiding the degradation of mental health that is common when a person undergoes changes related to training load, a new culture, and a new identity. This intervention described here is the first clear approach to implementing intervention-based models and clear communication in the Estonian Military Academy.

Intervention Goals, Target Groups, and Specific Theoretical Grounds

The resilience training intervention was introduced to cadets in a course named Applications of Psychology in the Defence Forces as an individual homework.

It is a second-year course in officer's education and covers different topics from the basics of military psychology, motivation, and leadership to psychological combat readiness (4 EAP). The author integrated this form of homework as it was not part of the coursework before. Thus, it is important to note that even though the article talks about intervention, the start of the program was dictated by the unique situation facing the world and lecturers in spring 2020. Readers are asked to accept the limitations of that time that contributed to a lack of well gathered quantitative data for the intervention results, as the focus was on supporting the cadets. Having said that, the intervention was based on theoretical grounds as mentioned in this chapter and the authors' expert knowledge.

This intervention was strongly affected by previous scientific research and programs in the military as well civilian and sports fields. In the military, we talk about performance, results, winning, and warrior mindset. High workload is expected and cherished. This also applies for the Estonian military community. The discussions of well-being and strengthening the mind will be guided by applying sports psychology instead of clinical psychology to help make the psychological preparations easier to comprehend.

Ward et al.[19] demonstrate the commonalities between military and sports domains, for example, both often require individuals to (a) perform in a complex and dynamic environment; (b) utilize a combination of perceptual, cognitive, and motor skills, (c) obtain a tactical advantage over their opponent; (d) act upon partial or incomplete information that evolves over time; (e) work both independently and as a team in an effective manner; and (f) operate under stressful circumstances. These aspects can and should be trained daily. Harms et al.[20] suggest that by being aware of the skills that appropriately characterize performance and identifying practices that help to improve performance, we can ensure that training is cost-effective and improves performance in the most expedient and efficient manner. For athletes, competitions are the ultimate test; for soldiers, the highest actual stress is experienced during missions. For both, the challenge is to achieve a mental state similar to that of an actual competition/mission to properly train the mental state.

Intervention Principles

People respond differently when experiencing stress and have distinctive motivations for getting through long uncomfortable and ambiguous periods, uncertain times, and seemingly endless piling of stress. Optimistic quotations such as "Stay motivated!" are not sufficient. What does work? Cigrang et al.[21] studied air force recruits and found that two sessions of stress-management

Built-in resilience intervention in academic military education

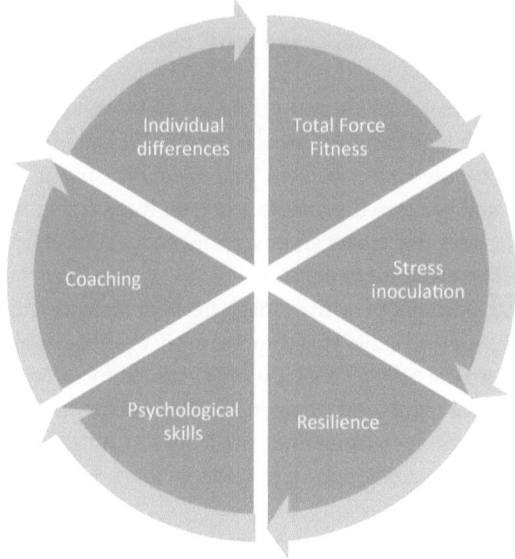

Figure 1: The six principles and concepts underlying the intervention.
This figure is original work created by Liisi Toom.

training did not make a difference for the mental state of at-risk trainees. However, Williams et al.[22] describe a study with navy recruits that involved an intervention designed to reduce emotional reactions and improve cognitive-behavioral coping strategies among at-risk recruits. Those who took weekly cognitive behavioral skills training were more likely to complete basic training compared to at-risk recruits without relevant training. The results suggests that recruits who receive constant support and obtain new evidence-based skills cope better with the tough demands of basic training. The intervention was based on six principles and concepts as depicted in Figure 1.

Total Force Fitness. It is important to take a holistic approach to life. For example, in military education, the focus might be on developing one's military identity, but other areas of life should not be ignored. A leader should be aware of what soldiers are going through in different parts of their lives, to better support them as needed.

1. **Stress inoculation** is like administering a vaccine: stress is introduced in small, manageable doses so that a person can adjust and learn vital coping skills. It is not only difficult but also potentially unethical to re-create true battle-field stress in a training. The best way to promote stress inoculation

as a means for mental strengthening is to use naturally stressful situations (like the COVID-19 pandemic) and teach cadets to develop healthy coping mechanisms.
2. **Resilience.** Knowing that mental resilience can be trained is a vital puzzle peace that can change the mindset of a person. We tend to believe that some people are naturally more resilient and mentally strong. This belief does not serve us well because it prevents a person from trying to find solutions. A leader who does not create an environment where followers can develop mental skills can cause more damage.
3. **Psychological skills** are present in many fields of military training, but rarely described as separate skills that can be used in other training fields. The author of this article believes that it is important to focus on motivation, goal setting, emotional regulation when engaged in military planning, leadership, and coping with the environmental challenges of warfare to strengthen the ability of military members to appropriately learn and apply these critical skills.
4. **Coaching.** Coaching is widely used in the military as a tool for leaders to improve followers' performance, cultivate self-motivation, and encourage decision-making. Teaching the basics of coaching to cadets helps them to develop this skill early in their career. It will support mutual understanding and improve the abilities of reflection and awareness within oneself and among each other.
5. **Individual differences.** Recognising different stressors and being open to understanding what works for each individual allows an intervention program to be flexible and help cadets to develop their toolkit according to individual needs. What this means with respect to the intervention process is that, despite a universal training of certain psychological skills, cadets must have the freedom to choose what they feel is most relevant for them in practice.

These six principles helped to create an intervention environment for growth mindset, cohesion, and self-awareness. These concepts have shown viability in previous research and will be in focus for further interventions and research.

Intervention Design

The intervention introduced to students was focused on achieving a performance goal by applying a variety of psychological skills (e.g., motivation, self-talk, relaxation). Ten students participated in the initial intervention (cadets of the navy and air forces). The intervention took place in March–April 2020 when the COVID-19 outbreak hit Estonia and the Academy adjusted with remote learning. It lasted for one month and included two short videos (on resilience and psychological skills), written materials, tracking with the HabitShare app, and a final report by the cadets. The second, improved intervention was introduced to 74 cadets (from the army, air force, and navy) in January–February 2021 and consisted of a single lecture on resilience and psychological skills, supportive lecture on identity and emotion regulation, a worksheet with instructions for setting goals, cadets coaching each other (pairs formed by the lecturer), questions for self-reflection, and a follow-up seminar.

The initial intervention was inspired by renowned military programs, like Battlemind, BOOT STRAP, and Mental Skills Training[23], using a wide range of exercises, including goal setting, self-talk, and mental training to enhance the mental and emotional parts of psychological functioning over a set period (e.g., 10 weeks for Mental Skills Training). During the course, similar techniques and goals were provided to cadets in a different design, time, and context. The theoretical part was short and conducted through video clips and e-seminars in a lecture format. The lecturer decided to apply the intervention when the COVID outbreak hit, to give the cadets specially focused activities that would also cover the *motivation* section in the curriculum for the course.

The specifics of our initial intervention were as follows:

1. A duration of one week to familiarize with the skills and figure out a goal and four weeks to practice.
2. During this time the students received approximately 45 minutes of lecturing on resilience and basic mental skills.
3. Students had to choose a goal (a physical achievement) and use a mental skill of their choice. They were given additional materials and links on breathing, mindfulness, relaxation, etc. and encouraged to find techniques suitable for them.
4. The cadets were asked to work in pairs as a coach and a coachee and were asked to meet three times. The meeting location and length was not fixed by the lecturer; the cadets were permitted to select these since the COVID outbreak influenced each student differently. The cadets received some literature on how to "coach," but no specific training or practice was possible.
5. After the intervention, the cadets submitted a reflective report. These reports demonstrated that the cadets understood the value of mental training and how it supports their well-being.

The **second intervention** applied all the previous steps, but with a few improvements and more time given for each activity. The initial group had fewer guidelines about the intervention due to its *ad hoc* nature. There was evidence that the results of the initial intervention were less clear, cooperation between the cadets as a coach and a coachee was not used to the fullest, and it was hard to understand how and to what extent the psychological skills were acquired and whether the cadets considered them useful. To improve these aspects, more specific guidelines were created in the form of a worksheet that covered the technical parts (how many times must a person meet with the coach/coachee), contents (which questions to ask a coachee, how to differentiate a result-goal from a psychological skill goal), etc. As this group received more lectures in the process, the lecturer could emphasize the importance of the intervention as well as ask for feedback and reply to questions that helped the cadets to clarify the assignment. The lecturer divided the group into coaching pairs by chance and provided a final report questionnaire for individual self-reflection. This helped the cadets to follow a clear goal while learning new useful skills and keeping autonomy over individual challenges. As a result, the depth of the intervention was visible from positive feedback, interest in participating, and the outcomes reflected in the reports that will be discussed later.

It is impossible to give exact length and content of each topic, lecture, and material because it was integrated with the general course. This was especially the case for the second intervention. Considering the significance of the integration it is important to draw out more clear content, length, and forms of measurement in the future to better apply the tools and integrate them more effectively. Should the reader be interested in more detail with respect to intervention topics, key teachings, mediums and actions taken by cadets, please contact this author for a more in depth table.

Results and Critique

Prior research has shown that psychological skills training has positive effects on the self-confidence, cohesion, and performance of athletes[24]. Ward et al.[25] noted that training improves when it is focused on key perceptual-cognitive skills, resulting in meaningful changes in the field. The results of the two interventions discussed in this article were measured only by the self-reflection reports of cadets. For the initial group, the report consisted of 5–7 open questions, while the second intervention was concluded with a self-reflection questionnaire that consisted of 18 questions (including "Please clarify further"). The reports mostly contained open questions, making it hard to give clear-cut and measurable

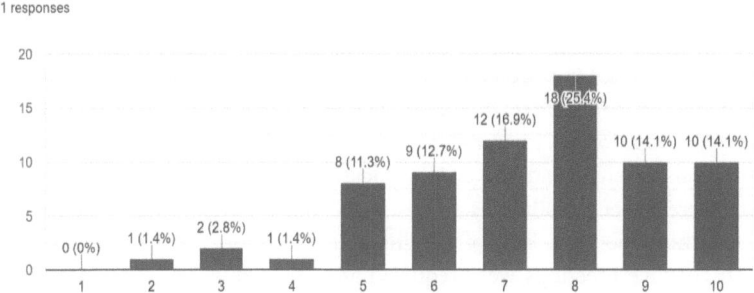

Figure 2: To what extent did this intervention help you enhance your chosen performance? 1-not at all/it hindered my performance – 10 – The benefit of the intervention was absolute.

This figure is original work created by Liisi Toom.

results. Our conclusions are thus based mostly on written comments and their interpretation and previous research results that are transferable to this intervention. 71 cadets filled in the questionnaire for the second intervention. 50.7 % of the cadets in second intervention had chosen "self-talk" as the skill to practice, followed by "motivation" (32.4 %) and various breathing techniques (21.4 %).

The objective of the intervention was to give the cadets a focus during a highly stressful time and provide realistic control over their psychological skills, resulting in better mental health, resilience, and self-efficacy, as well as improved physical health and performance. As illustrated in Figure 2, 59 cadets stated that the psychological skills exercise helped their performance. Both interventions helped to establish greater cohesion among students. Cadets reported being more motivated to take action. Here are some of the learning points cadets reflected: "I do sports regularly but I wanted to push my own boundaries with this exercise", "Each week it was easier and easier to work on my goal and skill", "I felt I needed to notice the good things more", "I learnt that you should not set a goal you can't reach with the time you have", "I noticed that when your coachee doesn't want to talk much and share information about him/herself, asking good questions helps him/her open up." All these learning points will be helpful to prepare also for potentially dangerous situations in their future careers as 73.2 % of the respondents answered that they could use the psychological skills learnt during the course in daily life.

There were considerable differences between the initial and the second intervention. The first big difference was using course mates to imitate a coach-coachee relationship, and the benefits it had for the cadets. Evident from the self-reflections and follow-up seminar discussions of the first intervention, the cadets did not see the opportunity or the value of communicating with a designated coach/coachee. The author believes that it was most likely because they were just forced to transfer to remote learning and were struggling to manage themselves and the communication among each other. As a result, there were fewer possibilities to stress the importance of the coach-coachee relationship due to a lack of seminars in between. For the second group, coach-coachee meetings were required and the group was physically present in the Academy, which made meeting and communicating with a designated course mate easier. 47.9 % of the students met 4 or more times (the minimum requirement was 3). They also received suggestions on how to coach and were provided with better explanations on how coaching a new person (not a friend or a course mate who you know well) is beneficial as a future military leader. As a result, the coach-coachee experience was highly valued. One person reported that the opportunity helped them to get better acquainted with course mates ("It is unbelievable that we have studied together, side by side, for half a year and still I find out that I do not know them or their way of thinking,"), understand individual differences, learn to motivate other people, and experience leadership in action.

The second difference was putting a focus on psychological skills rather than on a fixed goal. In the first intervention, having a goal to direct personal focus was more important and easier for the cadets. Learning stress-relieving skills during a stressful situation is always more difficult than learning them while in a relaxed state. Therefore, it made sense to focus on a controllable outcome and simultaneously practice psychological skills. That way, the intervention helped to raise awareness about the need for psychological skills, which is also an important outcome. In the second intervention, the stress factors were not so active, leaving more time and opportunities to focus on developing psychological skills. Leaders as well as subordinates should train psychological skills while not under acute stress yet engage in activities that focus on performance.

The third notable difference was understanding psychological skills and transferring them to other military training and leadership tasks. In a stressful situation, it is more about fight or flight, or handling the situation rather than focusing on its use in the future. The first group could use the assignment to enhance their current situation. In the second intervention, the group had several extra opportunities to transfer the mental skills to, for example, a shooting instruction course where cadets could detect and implement psychological skills (emotion regulation, motivation) to better instruct other in using the weaponry.

This added value to the skills and enhanced the opportunity to practice them in a different situation and environment, making it a realistic experience.

The results of the intervention are based on the self-reflective reports of the cadets and on the questionnaire administered to participants of the second intervention. Research on resilience programs supports the choices of the author for the interventions (which skills to develop, length of the intervention, expected outcomes), the course of action, and the reported benefits and results both groups enjoyed throughout the interventions. When applying the intervention in the future, proven measures and questionnaires could be added or developed. The intention with this program was to reduce stress, increase cohesion, and enhance the psychological skills for mental resilience, which means that the intervention could use some questionnaires to measure these issues.

There is one key difference in our intervention that had not been used in programs like BOOT STRAP: pairing the cadets to achieve maximum cohesion and awareness in individual differences and the fact that leaders must be able to coach a subordinate, instead of commanding them. As one respondent reflected: "Oftentimes we don't know who we will be working with in the future. Having the opportunity to test this out was interesting and I we made new friends". It would be interesting to explore this aspect further and see whether an intervention is more successful with or without this aspect, or whether a group discussion could be more effective. Also, increasing the skills and significance on coaching and team activities could be a valuable learning point while helping to reduce stress, promoting mental health and teamwork, and achieving a better understanding on individual differences and needs. A cadet reflected that "the coach had a very big role to play when I was doubting my own values".

A big issue with this intervention was the ability to check the skills of the cadets and the level of applying them. Smaller groups and group-coaching or additional educational videos on different psychological skills could help to prepare and teach the necessary skills more effectively.

Further investigation should explore how and what exactly is affected with the intervention, adding clearly appropriate and quantifiable questionnaires to examine the actual change in performance, mental health, and resilience.

There is also a need to determine whether all employees of a military organization should require this kind of education. We argue that it is necessary for two reasons. First, each member should be able to support and understand the demands put on the soldiers and help them at any time, adding benefits to the work carried out by psychologists and designated professionals. If we look at well-being from a positive psychology perspective, a person does not need to be a psychologist to support resilience. Second, we believe that stress is part of everyday life and, as employees, we would be able to support our people and, therefore, increase the well-being, coping strategies, and resilience in the general society. Next, we will discuss further implementation of this intervention.

General Use for the Resilience Intervention Program Post COVID-19

Although we are hopeful that COVID-19 will remain a one-of-a-kind situation, stressful circumstances will still present themselves during peacetime in different ways. COVID-19 can prove to be a turning point for supporting well-being: positive psychology programs should become a norm at every level of military preparation, not just mission readiness or resolving clinical illnesses caused by failure to cope with the occupational demands of the military or personal life challenges. An integration of similar intervention programs could, therefore, support general well-being, resilience, and better life and leadership skills within the whole organization. Subsequently, there is a list of possibilities for using the learning points of the intervention in other organizations:

1. Stress inoculation is a tool for teaching cadets, soldiers, and military leaders to notice and reflect on mundane stressful situations and quickly respond with either additional training or coaching; it can also be used for aiding combat training and preparations. Leaders, subordinates, and the whole community will obtain terms and concepts to use, reversing the effects of stigmas and supporting healthy performance.
2. Focusing on positive outcomes and positive psychology provides a better tool for managing either a temporary or permanent transition from military to civilian life. Obtaining these skills early will hasten the process because the soldiers would already be one step ahead.
3. Interventions are applicable to enhanced sports performance, general combat training or specific shooting exercises, giving added value and deeper understanding, and promoting better communication among teammates. It would make training more autonomous for soldiers and mitigate the workload of instructors.
4. Learning coaching and reflection from early on can give a person the necessary empathy to quickly generate trust and leave space for a person, increasing cohesion within a team in the long run.
5. Each organization should have materials on general well-being, resilience, and psychological skills, and their potential appliance. Sharing such techniques in a video, brochure, leaflet, poster, or other mediums suitable

for an organization is a gradual and cost-efficient way to promote well-being and prepare workforce for stressful situations. Privacy is essential in these subjects and an opportunity to self-educate helps to prevent many issues.
6. The best way for a leader to understand the value of psychological skills and resilience training is to get personally involved. When leaders learn the skills and understand the mechanics, their influence will be that much stronger, continuously creating a stigma-free environment where people are able to thrive physically and mentally.

In summary, the elements of this program are a necessary educational and training basis for many organizations, teams, and professionals. We will investigate developing this program further at the Estonian Military Academy through possibly:

(a) Adding resilience training to the basic psychology course as a permanent part.
(b) Creating an open course for leading positions.
(c) Developing military leadership core competence model to include awareness building, coaching, psychological skills and positive performance environment creation skills which all could be part of training the core competencies of military leaders.[26]
(d) Communicating to raise awareness to support employees before, during, and after a stressful situation both individually and as a team.
(e) Researching leadership performance and mental training. We are looking for more scientific ways to understand the impact of and a need for this program.

The primary focus in leadership training could be on enhancing individual positive coping mechanisms, satisfaction from work, and self-efficacy to inspire a person to continue a military leadership career and become a positive role model. This would have an impact on the values and image of soldiers within the military culture as well as in society: ***mental strength comes first***. The author of this chapter believes that it can have a great impact on how military service is perceived and give cadets and future military leaders the tools to influence both mental and physical fitness in the wider society.

Communicating Mental Health in the Military

There is one topic that should be considered when integrating resilience programs for military personnel activities to make sure the programs are more accepted, valued, and implemented within and outside an organization. It is commonly believed that mental health refers to illness and should, thus, be partly or entirely excluded from military education. This perception does not suit either modern leadership or mental performance training in high-performance professions. In the military, positive psychology and sport psychology principles are more appropriate for supporting cadets and soldiers and the military culture. First, the military takes pride in selecting and recruiting the best, which means that it is also the source for many mental challenges. The right thing to do is to support the personnel in strengthening their mind as well as their bodies. Second, individual performance depends on perception, beliefs, and the ability of self-control. This is directly associated with psychological skills; unfortunately, this subject is not sufficiently covered in general military education. Third, discussions over stress and mental health should be more directed to 'normality': achieving the best life, enjoyment of work, and maximizing individual potential. Failures, challenges, and hardships are physically and mentally normal. It is comparable with supercompensation (in sport performance training) due to similarities with mental training.

This author suggests that mental health, well-being, mental performance, and other similar concepts should be integrated in military education and training at all levels to avoid stigmatization and to prepare military leaders as role models for mental preparation and well-being in the society at large. A clear distinction should be made between illness, diagnosis, and performance-related mental strength. Messages and displays that support mental well-being should stand beside posters of tanks and guns. If we change communication, less resources will be spent on minimizing stress, retention and mental illnesses caused by stigmas and fear of failure; we would enable the military to really use the skillful and motivated people they initially recruited. This type of environment helps to build organizational reputation and transfer the skills and mindset learnt in the military to civilian life since the military personnel can go on to educate their immediate communities and provide an example for the wider society.

Conclusion

Resilience and psychological skills training can be more generally integrated into the life, education, and military training of soldiers. As educators and/or leaders, we can use contemporary events to illustrate the use of mental aspects during peacetime while saving time and effort in preparing for wartime or in specific trainings. This chapter focused on an intervention that combined naturally stressful circumstances with academic studies to teach cadets the mental skills that they could immediately apply. This experience opened a discussion amongst cadets and the Estonian Military Academy personnel to further use psychology for the well-being and performance enhancement of cadets in several contexts. The author hopes that this type of integration would help to fight against the stigmas of psychology and open the military domain to a more consistent work on well-being and resilience amongst cadets, military leaders, and involved institutions. Mental training is a natural resource for creating mentally fit soldiers on every level. The military community has a possibility to inspire the wider public to value resilience and mental strength and be a role model on how to create stronger cohesion, sense of self-control, and more resilience towards naturally occurring adversities as well as challenges we face as professionals, individuals, parents, partners and leaders and teachers.

Notes

1 McGraw, L., Pickering, M. A., Ohlson, C.; Hammermeister, J, "The influence of mental skills on motivation and psychosocial characteristics." *Military Medicine*, 177, no. 1, (2012): 77–84.
2 Sequelae is most often used in medicine to refer to a condition that is a consequence of a previous injury or disease. It is also appropriate for psychological conditions.
3 The term "onboarding" is widely used in Human Resources in reference to activities carried out with new employees to help them to adapt to new work.
4 L.L. Hourani, T.V. Williams, and A.M. Kress, "Stress, mental health, and job performance among active duty military personnel: Findings from the 2002 department of Defense Health-Related Behaviors Survey." *Military Medicine*, 171, no. 9 (2006): 849–856.
5 Ibid.
6 Ibid.
7 R.A.Williams, B.M. Hagerty, S.M. Yousha, J. Harrocks, K.S. Hoyle, and D. Liu, "Psychosocial effects of the BOOT STRAP intervention in Navy recruits." *Military Medicine*, 169, (2004): 814–820.
8 E. Mõts, *"Eesti Kaitseväe Maaväe Lahingutegevuse alused,"* 2019.
9 C. Cipriano, G. Rappolt-Schlichtmann, and M. Brackett. *Supporting school community wellness with Social and Emotional Learning (SEL) during and after a pandemic* (2020). Pennsylvania: Pennsylvania State University, Issue brief.
10 Ibid.
11 Ibid.
12 A. Williams, B.M. Hagerty, S.M. Yousha, J. Harrocks, K.S. Hoyle, and D. Liu, "Psychosocial effects of the BOOT STRAP intervention in Navy recruits." *Military Medicine*, Vol. 169 (10) (2004): 814-820.
13 L.L. Hourani, T.V. Williams, and A.M. Kress, "Stress, mental health, and job performance among active duty military personnel: Findings from the 2002

Department of Defense Health-Related Behaviors Survey." *Military Medicine*, Vol. 171 (9) (2006): 849-856.
14 J.A. Martínez-Sánchez, "Psychological intervention in the Spanish military deployed on international operations," *Psicothema*, 26, no. 2 (2014): 193-199.
15 P. Lester, P. Harms, M. Herian, D. Krasikova, S. Beal. *The comprehensive soldier fitness program evaluation. Report 3: Longitudinal analysis of the impact of master resilience training on self-reported resilience and psychological health data* (2011). Nebraska: Public Policy Center, University of Nebraska.
16 C.S. Dweck. *Mindset: The new psychology of success* (2006). USA: Random House.
17 Angela Duckworth. *Grit: Why passion and resilience are the secrets to success* (2017). London: Vermilion.
18 V. Bowles, L.D. Pollock, M. Moore, S.M. Wadsworth, C. Cato, J.W. Dekle, S. Wei Meyer, A. Shriver, B. Mueller, M. Stephens, D.A. Seidler, J. Sheldon, J. Picano, W. Finch, R. Morales, S. Blochberger, M.E. Kleiman, D. Thompson, and M.J. Bates, "Total force fitness: The military family fitness model." *Military Medicine*, 180, no. 3 (2015): 246-258.
19 P. Ward, D. Farrow, K.R. Harris, A.M. Williams, D.W. Eccles, and K.A. Ericsson, "Training perceptual-cognitive skills: Can sport psychology research inform military decision training?" *Military Psychology*, 20 (2008): 71-102.
20 P.D. Harms, D.V. Krasikova, A.J. Vanhove, M.N. Herian, and P.B. Lester, "Stress and emotional well-being in military organizations." *Research in Occupational Stress and Well Being*, 11 (2013): 103-132.
21 J.A. Cigrang, S.L. Todd, and E.G. Carbone, "Stress management training for military trainees returned to duty after a mental health evaluation: Effect on graduation rates." *Journal of Occupational Health Psychology*, 5, no. 1 (2000): 48-55.
22 R.A. Williams, B.M. Hagerty, S.M. Yousha, J. Harrocks, K.S. Hoyle, and D. Liu, "*Psychosocial effects of the BOOT STRAP intervention in Navy recruits.*"
23 Harms et al., "*Stress and emotional well-being in military organizations.*"
24 Harms et al., "*Stress and emotional well-being in military organizations*".
25 P. Ward, D. Farrow, K.R. Harris, A.M. Williams, D.W. Eccles, and K.A. Ericsson, "*Training perceptual-cognitive skills: Can sport psychology research inform military decision training?*".
26 Ü. Säälik, A. Ermus, I. Männamaa, L. Toom, and A. Kasemaa, "Kaitseväelise juhi pädevusmudel." *Sõjateadlane* (*Estonian Journal of Military Studies*), 14 (2020): 11-38.

Walter L. Giusti and Gayle Sherwell

From Military to Government Sector Applications: Developing an Organizational Climate Framework to Promote Psychological Health, Well-being and Positive Work Behaviors[1]

Introduction

The large-scale risk profiling and management of personnel mental health and well-being in military and government organizations is often undertaken in the context of psychological and organizational climate (climate) monitoring. Climate profiling tools are designed to measure personnel and employee perceptions of what it feels like to work in the organization, the workplace conditions, how the conditions affect their health and well-being, how committed they are to their job and how effectively they and their unit operate on a daily basis.[1]

The purpose of undertaking climate profiles is to provide comprehensive and actionable information to senior decision makers on the health, well-being and morale of personnel and their units, workplace conditions that impact on them, as well as their effect on work outcomes. This information can then be used to decide on the best course of action to take for remediating identified risks and improving the climate of an organizational unit or agency.

Much of the early pioneering work and large-scale application of psychological and organizational climate profiling had been undertaken by military establishments, who had long recognized that the health, well-being, and morale of their soldiers were necessary conditions for the effectiveness of their unit. Consequently, forward-looking military organizations have made great progress in developing evidence-based systems and tools for the large-scale monitoring of personnel mental health and well-being aided by the principles and methods of social and behavioral science and psychometrics.

---

1 The views and opinions expressed in this chapter are those of the authors and do not necessarily represent the official policy or position of any Australian government agency or of the Australian Department of Defence.

Following this example, we developed the Justice Working Well Framework (JWWF) as a research and development (R&D) project for a large Australian government justice agency (the Agency). The aim of the project was to develop and trial a comprehensive and 'evidence-based' surveillance, profiling, and management system capable of identifying psychological and organizational climate risks across units of the Agency. The system sought to provide senior leaders with actionable information for initiating remedial interventions to mitigate identified risks and to promote a positive work environment. We also focused on tailoring the Framework to satisfy the specific occupational health and safety (OH&S) needs of the Agency and to contribute to its OH&S strategy. At the same time we aimed to address the current limitations of climate monitoring systems commonly used in the government human resource sector by integrating advances in knowledge in the field.

In this chapter, we begin by outlining the rationale for the development of OH&S monitoring tools in the Australian government sector. We then discuss the development and operation of the major components of the JWWF, including (a) lessons learnt and key contributions from an Australian Defence Force (ADF) climate surveillance system – the ADF Profile of Unit Leadership, Satisfaction and Effectiveness (ADF PULSE), which informed the construction of the conceptual model and metrics of our profiling system; and (b) additional features and protocols that we introduced to the Framework, with an explanation of the rational for these innovations. This includes an outline of key innovations to the Framework, based on the principles of action-research modelling, which enables it to function as an iterative system for promoting continuous improvement in psychological and organizational climate. In line with the theme of this volume, we also highlight the central role that employee health and well-being play in the Framework and conclude this chapter with suggestions for further work required to make it more effective.

Background: Rationale for Development of the JWWF

The military establishment has long recognized the importance of the health and morale of soldiers to their unit's cohesion and functioning, both in barracks and in combat. The application of behavioural and social science methods to the large scale surveillance of soldiers attitudes, well-being and morale, and the effect of these factors on their units effectiveness, has its origins during the Second World War in the work of Samuel A. Stouffer (a University of Chicago and Harvard sociology professor) and his colleagues, which resulted in the publication of a number of seminal reference volumes on the American soldier.[2] This pioneering work on the attitudes and morale of US soldiers influenced the decision making of pivotal military figures such as Generals Eisenhower and Marshall. But also critical to the adoption of these new science-based methods was the support of senior government figures such as General Frederick H. Osborn (Chief of the Army's Information and Education Division), who promoted Stouffer's work in the US government, and US President Franklin D. Roosevelt, who understood the value of this new evidence-based thinking for improving the effectiveness of military and government organizations.[3]

These pioneering developments, together with parallel advances in industrial and organizational psychology, psychometrics, and social survey methods, served as the foundation for the large-scale application of contemporary psychological and organizational climate profiling tools in the military. Notable for their strong evidence-base, are two related military climate surveillance tools – the Canadian Forces Unit Morale Profile (CF UMP) and the Australian ADF PULSE. The later has its origins in an exchange program between the Australian Department of Defence and the Canadian Department of National Defence in the late 1990's and early 2000's.[4] A major strength of both these climate profiling systems is that they were developed from a strong evidence-base, in line with the principles of organizational psychology and psychometrics. In addition, the agencies responsible for their administration continue to regularly review and refine the profiling metrics of the CF UMP[5] and ADF PULSE[6] according to changes in organizational needs, and advances in best practice in psychometric knowledge.

Following the military example, government organizations have increasingly committed to undertaking organizational climate surveillance and implementing employee health, well-being and safety initiatives in order to promote the effective functioning of their workforce. These have also been driven by several economic, legal, and technical factors, namely:

- the rising cost to organizations of workplace accidents and poor health of staff, that result in higher absenteeism, employee turnover and health insurance claims;
- advances in industrial and workplace relations that promote the social responsibility of employers towards their employees, in conjunction with the introduction of workplace legislation that supports the right of employees to safe and healthy workplaces;
- the growth of knowledge on the critical role that employee health, well-being and morale play in work outcomes, organizational capability, and effectiveness; and

- advances in the development of evidence-based methods for the surveillance and promotion of employee health and well-being in the workplace.

First, a significant proportion of human resource costs to organizations are through work insurance claims, absenteeism and staff turnover. National surveys have indicated that the health of the workforce in Western countries is declining and that mental injury claims are on the increase. In Australia, for example, the cost of employee absenteeism to the economy has been estimated at around 7 billion dollars a year, and the loss world-wide has been estimated to be in the trillions of dollars.[7]

Second, developments in the field of industrial relations and the introduction of workplace legislation that require organizations to ensure that workplaces are healthy and safe have prompted employers to demonstrate greater social responsibility towards their employees. For example, the Australian State of Victoria's Occupational Health and Safety Act (2004) requires an employer, so far as reasonably practicable, to provide and maintain a working environment that is safe, free from harm and without risk to the health of employees – both physically and psychologically (Section 21 (1)).[8] In line with the legislation, which is legally enforceable, government agencies such as Safe Work Australia are responsible for providing guidance on employer's duty of care regarding psychological health and safety in the workplace.[9] Industry and professional associations have also contributed to this effort – the Australian Psychological Society (APS), for example, has developed a national initiative to assess and entitle businesses and organizations to use the designation of 'APS Certified 'Psychologically Healthy Workplace' in their staff recruitment and advertising campaigns.[10]

Third, these developments have been supported by advances in knowledge in the fields of applied, industrial and organizational psychology that have well demonstrated the advantages of ensuring ethical, healthy and safe work environments for employees.[11] In particular, research on psychological and organizational climate in various workplace settings has provided strong evidence for the relationship between workplace conditions and employee health and well-being, on the one hand, and work outcomes such as the level of work engagement, productivity, performance and organizational effectiveness, on the other.[12]

In response to these developments, and in order to demonstrate that they are meeting their duty of care towards their employees, Australian government sector agencies have implemented comprehensive workplace health and safety policies, strategies and initiatives.[13] This includes the large-scale application of surveys for the purpose of periodically gauging the work attitudes, morale, health and well-being of their employees. In Australia these surveys are normally administrated periodically through central government agencies, such as federal and state public sector commissions.[14]

However, most employee surveillance tools currently used by government agencies tend to suffer from numerous limitations. Fogarty and Machin[15] have identified several key challenges facing the application of these tools, namely that:

- Requests for climate surveys are not driven by clear modes and protocols for how the findings can improve organizational outcomes.
- The resulting questionnaires tend to lack established psychometric properties and benchmarking.
- The tools generally lack established protocols for communicating results to staff and how to use findings for strategic planning and initiating actions for organizational improvement, and
- The interpretation of survey findings is usually 'the exclusive province of experts, without whose help little can be done by the organization itself'.

Given the important role that climate profiling serves in informing high level decision making on matters of organizational risk and effectiveness, forward thinking government agencies have begun to address these challenges by investing resources towards developing more evidence-based, action-orientated, and user-friendly tools and systems. Notable is the work undertaken by Fogarty and Machin for the Organisational Improvement Unit of the Queensland State Government's Health Department (Australia).[16]

To contribute to these efforts, and to address the need for a robust OH&S monitoring system in the government justice sector, we undertook an R&D project to develop the Justice Working Well Framework as an evidence-based, user friendly and research-action orientated system for the monitoring and management of psychological and organizational climate risk factors.[17] Reference to Fogarty and Machin's key criteria meant that the Framework should ideally incorporate: (a) a model and associated protocols for how the findings could be used to identify risks and improve organizational outcomes through appropriate action; (b) an evidence-based climate profiling tool, here referred to as the Justice Working Well Profile (JWWP), with good psychometric properties and benchmarking capability; (c) protocols for communicating results to decision makers, and for making use of findings to initiate actions and remedial interventions to improve climate; and (d) a user-friendly system where the core profiling function can be applied and interpreted by 'generalist' government sector staff.

In the following sections, we present an outline of the various innovations and components that have been incorporated into the JWWF to satisfy these criteria.

Constructing the JWWF as a Modular Action-Research System

To address the limitations of current psychological and organizational climate frameworks, we built into the JWWF multiple components and processes that work together to systematically identify and reduce climate risk to the Agency. As a starting point, we sought to establish a rigorous psychometric foundation for the Framework's profiling component without needing to 'reinvent the wheel'. For this purpose, we modeled a number of features of the Framework on those of the ADF PULSE. These include the conceptual model of psychological and organizational climate, core climate measures and scales, and protocols for their continuous improvement. And drawing on Fogarty and Machin's recommendations on the utility of undertaking relational modelling in conjunction with climate profiling, we proposed the inclusion of this feature into the Framework as a value-added optional capability for future development.[18]

After identifying the Agency's specific OH&S needs, we incorporated additional features into the Framework to address these. These included: (a) additional scales of Workplace Safety Climate, Occupational Fatigue, and Work Engagement; (b) a toolkit to guide recommendations and initiate actions designed to reduce identified areas of climate risk – this was developed by matching the Agency's exiting OH&S and training programs to the climate facets measured by the JWWP; and (c) an evaluation protocol to provide feedback on the effectiveness and impact of the remedial actions that are taken. Individually, these additional components are not new to the profiling and management of personnel mental health and well-being in an occupational setting. However, what is novel is the integration of these functions into a single framework that enables it to function as an iterative 'action-research' system for the continuous improvement of psychological and organizational climate.

Action-research is a school of thought, set of principles and methods that have their origins in the work of the social psychologist Kurt Lewin.[19] Models of action-research typically seek transformative change through a predefined cycle of data driven enquiry, analysis, initiation of problem-solving actions, critical evaluation of outcomes of the actions, and the identification of any new problem that are emerging. Compared to linear processes of change management, the advantage of this iterative process of learning, evaluation, and revision, is that with every application cycle it leads to increasingly better outcomes and the continuous improvement of a system.

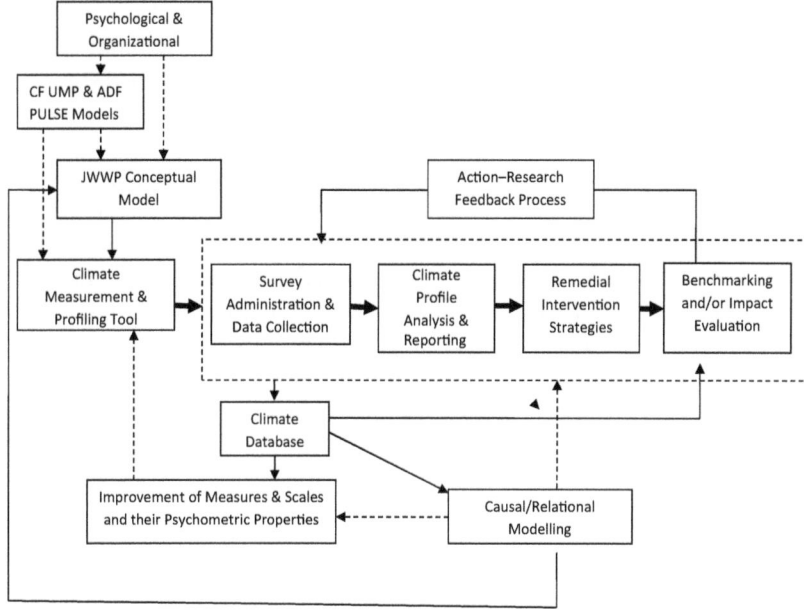

Figure 1: Operational model of the psychological & organizational climate integrated framework.

Figure 1 illustrates the way that the components of the Justice Working Well Framework interact as an action-research system. How each contributes to the Framework is discussed in the following sections.

The Conceptual Building Blocks of Climate: Domains and Facets of the Justice Working Well Framework

Building the conceptual model of core domains and facets of psychological and organizational climate was the first step in the development of the Framework, as it forms the basis for the measures, scales and profiling component of the system (described in the next section).[20] In building the model we needed to first distinguish organizational climate from organizational culture. The former refers to employee's cognitive appraisals of what it feels like to be a member of, and work in, an organization. In contrast, culture covers a set of values, customs, rituals and behaviors that evolved as part of the organizations adaptive process.[21] Although these concepts are related, anthropologists such as Schein[22] have argued that the measurement of culture should not be limited to survey questionnaires, a standard tool used by organizational and industrial psychologists to measure climate, but also requires the use of ethnographic methods and close observational engagement with the organization. As the topic of our paper is climate, here we focus exclusively on the former.

In constructing the conceptual model of the JWWP, we used both theoretical and practical criteria to identify the core categories and facets of climate relevant to the Agency. The ADF PULSE served as a good starting point,[23] as it is a tried and tested psychological and organizational climate profiling tool which is well supported by the research literature.[24] We tailored this model to the specific needs of the Justice Agency through consultation with senior leaders in the OH&S area and reference to the organizations OH&S strategy. Consequently, we modified several facets, and introduced additional ones, such as those of Occupational Fatigue and Work Engagement, to supplement the three core domains in our model – (1) Workplace Facets, (2) Employee Attitudes and Psychological Health and Well-being, and (3) Work Outcomes. The structure of the JWWP model is illustrated in Figure 2.

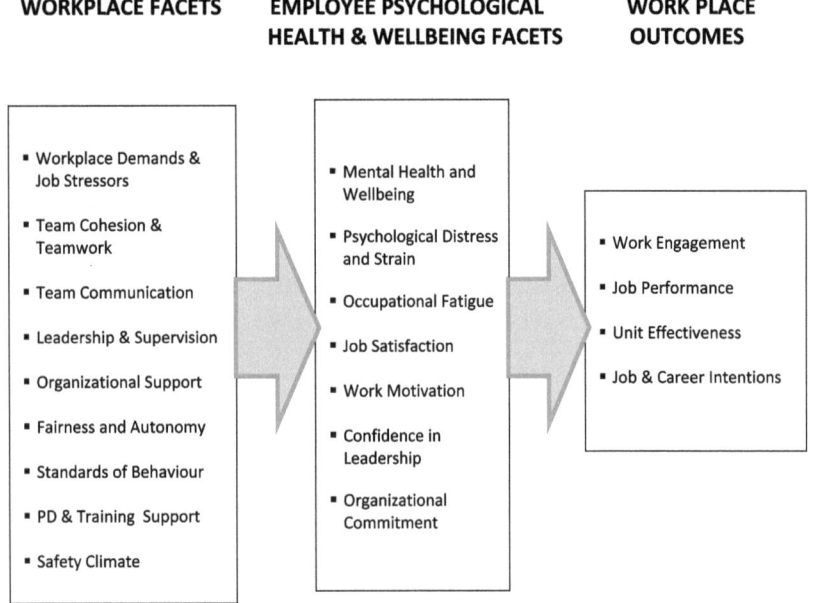

Figure 2: Conceptual model of JWWP's psychological & organizational climate domains and facets.

The categories and related facets identified in the model provide the basis for the development of JWWP measures and scales (discussed in the following section). It is noteworthy that facets of employee mental health and well-being play a central and mediating role in the model, linking other domains of workplace climate to those of work outcomes and effectiveness, a relationship that is well supported by the research literature.[25] However, while our illustration suggests a simple linear association between these domains and their facets, the actual relationships are more complex. The utility of applying advanced techniques of multivariate analysis to examine the complex web of interactions between climate domains and factors identified in the JWWP framework is highlighted in the section on causal and relational modelling, below.

Measurement and Scale Development

Fogarty and Machin[26] point out that too few employee attitude climate monitoring tools used by organizations are sufficiently grounded in evidence-based methods from the social and behavioral sciences,[27] or psychometric principles of scale development.[28] Consequently, most of these tools have untested and unknown validity and reliability. Therefore, in constructing the JWWP – the profiling function of the Justice Working Well Framework – we focused on adopting and further developing measures based on sound psychometric principles in order to ensure confidence in the outcomes of our climate profiles.

We began by operationalizing the conceptual model of the JWWP as scales and sub-scales to enable the measurement of its core facets. This process was facilitated by a knowledge sharing agreement between our Agency and the Mental Health, Psychology and Rehabilitation Branch of the Australian DoD's Joint Health Command, which enabled us to share lessons learnt from the development of the ADF PULSE. In order to tailor the JWWP to the specific requirements of the Justice Agency, we adopted core scales from the PULSE, including measures of employee mental health and well-being, such as job stress, motivation, psychological distress and general psychological health, as well as introduced a number of additional scales to our profiling toolkit. These included the modification of the Employee Professional Conduct and Standards of Behaviour scale (for the Workplace Facets domain), and the introduction of new scales on Employee Work Engagement (for the Workplace Outcomes domain), and Occupational Fatigue (for the Psychological Health and Well-being domain), the latter being a modification of the Fatigue-at-Work Scale (FAWS) which was originally constructed by Giusti[29] for the purpose of measuring post-operational fatigue as an indicator of health and well-being of ADF personnel returning to Australia after deployment to the Middle East area of operations. The development of the FAWS for specific military application drew considerably on the content and structure of the Multidimensional Fatigue Scale (MFI), with permission from the original authors, as it was a well tried and tested measure, with excellent psychometric properties, of both acute and chronic fatigue across a number of dimensions.[30]

Additionally, the short self-report version of the Utrecht Work Engagement Scale (UWES-9) was introduced to the JWWP toolkit, as it is a psychometrically robust measure of the extent to which employees engage positively with their work – i.e. 'have a sense of energetic and effective connection with their work activities and see themselves able to deal well with the demands of their jobs'.[31] The UWES-9 is a two-factor scale measuring Engagement (i.e. vigor, dedication, absorption and professional efficacy) as well as its negative antipode, Burnout (i.e. exhaustion and cynicism).[32] Measures of work engagement have been extensively researched and shown to be positively related to employee work outcomes such as performance and productivity.[33] This scale was introduced to supplement existing measures of work outcomes such as the self-assessed Individual Job Performance scale that was adopted from the PULSE and CF UMP, as the latter was more vulnerable to subjective social desirability influences.

A concern when building the JWWP was that climate survey questionnaires typically tend to be very large, with the number and length of scales contributing to their overall size. While larger scales tend to be more psychometrically robust, the use of shorter scales (composed of fewer items) reduce participant's experience of survey overload and survey fatigue, and therefore contribute to achieving better response rates. This is more of a concern when conducting surveys of government sector employees than of military personnel, because achieving adequate response rates is more challenging where employee participation is voluntary and undertaken electronically (this issue is discussed further in the next section on survey administration and data collection). So, a major challenge in constructing the JWWP was to develop a comprehensive climate profiling tool that was not excessively long, while at the same time taking care to maintain the psychometric integrity of its measures.

Therefore, when developing the initial JWWP prototype, we sought to use shorter but still psychometrically robust scales. This involved adopting short versions of existing scales, such as: the nine-item UWES-9 instead of the seventeen-item UWES-17[34] (mentioned above); the Kessler six-item K-6 instead of the ten-item K-10 (to measure non-specific psychological distress in terms of non-clinical depression and anxiety)[35] and the nine-item FAWS,[36] a modified and shorter version of the self-report twenty-item Multidimensional Fatigue Scale.[37]

Following the practice of the CF UMP and ADF PULSE for periodically examining and revising measures and scales,[38] after the first application of the JWWP, we sought to further streamline and reduce the size of the questionnaires. As well as ensuring the continuous improvement of the measures, this practice also facilitates their adaptation to the changing needs of the organization. The process includes the removal of poorly performing or redundant scales and subscales which are shown to contribute little to the overall climate of the organization, as well as the elimination of items from scales that do not contribute to their reliability, while at the same time taking care to maintain the integrity of the instrument. With the JWWP, care was taken particularly when revising measures of employee health and well-being, as these tend to be more susceptible to social desirability responses due to the stigma attached to issues of mental health in the workplace.

After our initial revisions of the core measures of the JWWP, we re-tested the psychometric properties of the scales, their factor structure and reliability using data collected from the initial pilot study ($N = 98$).[39] Table 1 presents a summary of reliability outcomes for the core JWWP scales (but not their sub-scales) in terms of the Cronbach's alpha coefficient, and compares these to like scales of the ADF PULSE[40] and CF UMP.[41] The generally agreed lower limit for Cronbach's alpha is 0.7, although as low as 0.6 may be tolerated for exploratory research or for scales with fewer items where the reliability requirements are less stringent. While reliability estimates of 0.8 and above are regarded as moderate to good, and 0.9 and above as high.[42] On average, all three profiling tools demonstrate good levels of scale reliability, particularly given their use of relatively short scales. Note that despite large differences in sample sizes, similar levels of reliability were still achieved with comparable climate scales across the three different populations – Canadian Forces personnel (N=4885), Australian Army personnel (N=968), and Justice Agency employees (N=98)[43]– demonstrating the consistency and applicability of these measures across different national and occupational populations.

Table 1: Average Scale Reliabilities (Cronbach's Alphas) for CF UMP, ADF PULSE and JWWP

Climate Scales	Organizational Climate Tool		
	CF UMP (N=4885)	ADF PULSE (N=968)	JWWP (N=98)
DOMAIN 1 - WORKPLACE FACTORS	0.82	0.80	0.84
Job Role Stressors	0.8	0.78	0.77
Teamwork & Team Cohesion	0.76	0.69	0.75
Supervisor Performance	*	0.93	0.94
Satisfaction with Communication	0.8	0.76	0.89
Perceived Organizational Support	0.94	0.87	0.90
Professional Conduct/Behaviour at Work	*	*	0.82
Perceived Organizational Fairness	0.95	0.77	0.88
Employee Autonomy	0.71	0.79	0.79
Workplace Safety Climate	*	*	0.92
Professional Development and Training Support	0.78	*	0.74
DOMAIN 2 - PERSONNEL MENTAL HEALTH, WELLBEING & MORALE	0.83	0.72	0.78
Psychological Distress (K10)	0.93	0.93	0.85
Work Motivation	*	0.74	0.72
Confidence in Leadership	0.72	0.56	0.75
Organizational Commitment*	0.81	0.65	0.77
Job Satisfaction	0.85	*	**
Fatigue-at-Work	*	*	0.82
DOMAIN 3 – WORK OUTCOMES	0.80	0.81	0.85
Work Engagement	***	***	0.78
Job Performance	0.79	0.81	0.89
Career/Job Intentions	0.81	*	0.89
AVERAGE SCALE RELIABILITY	0.82	0.77	0.83

* The scale or subscales of this construct are not structurally similar to that of the JWWP, and therefore not comparable.
** This is a single item scale and therefore not able to return a Cronbach's alpha co-efficient.
*** This climate tool does not contain a like scale with that of the JWWP for comparative purposes.

Survey Administration and Data Collection

The operational phase of psychological and organizational climate profiling begins with data collection. For the ADF PULSE, climate surveys are initiated by Unit Commanders putting in a request to the Divisions Psychology Support

Section (PSS), who then liaise with the Directorate of Occupational Psychology and Health Analysis (DOPHA) in undertaking a profile of the Unit. DOPHA then coordinates the process of survey administration, data analyses and reporting of outcomes, that are used by PSS to brief the CO's. Like the PULSE, the JWWP was initially designed to be applied by the Agencies Health and Safety Unit in response to requests from senior leaders for a climate profile of their units or divisions.

In addition, the Framework was also designed to enable the periodic climate surveillance of Agency units, following the practice of Australian Federal and State public service commissions that conduct large scale government employee attitude surveys on regular basis.[44] Periodic surveys have the advantage of providing more systemic monitoring of climate by facilitating benchmarking and comparison across time and across units. Hence, periodic profiling remains an option for the Agency, especially given the automated administration and delivery capability of the JWWF.

For both JWWP and PULSE, data collection involves the delivery of climate survey questionnaires, along with instructions, to prospective participants of organizational units selected to undergo profiling. There are, however, major differences in the methods of delivery between the two. At the time of development of the former, PULSE survey questionnaires were administered by Army psychologists in 'pencil and paper' format to ADF personnel in barracks as a group exercise. Completed questionnaires were then scanned and coded, and the data entered into electronic spreadsheets (such as EXCEL and SPSS) before undergoing analysis by DOPHA. In contrast, to meet our organizations requirements, we automated the JWWP survey process by embedding it in the survey function of the Agency's Personnel Performance Management System (PPMS). In this way we could deliver questionnaires electronically to staff and collate raw response data using a dedicated database.

The advantage of the JWWF automated survey administration and data collation system is that it requires fewer resources to run, and it can also be programed in advance to run periodically. However, the PULSE face-to-face method of delivery is able to achieve participation rates of almost 100 percent, compared to around 30 percent for the JWWP (which is normal for government employee surveys of this sort). The exceptionally high rate of participation achieved by the PULSE can be attributed to a number of reasons relating to the military culture and *modus operandi*. These include the more personal administration of questionnaires by unit psychology officers and their delivery on site in barracks and within a scheduled timeframe, the stronger requirement for participation of military personnel in surveys, and the more direct involvement of Unit Commanders in facilitating the process.

For the JWWP, like all climate surveys, it is important to achieve reasonable response rates and sample sizes in order to increase the statistical power of the analyses,[45] and also to enable the 'slicing and dicing' of data for drill-down comparisons between demographic categories. To achieve this, effective strategies need to be applied to improve the response rates of government employees, including those that reduce the risk of 'survey overload' and 'survey fatigue', which are major causes of low employee participation.

Analyzing and Reporting Climate Profiles

The analysis and reporting of large amounts of survey data that cover many facets and sub-facets of psychological and organizational climate can be very resource and time intensive. In order to enable quicker turnaround times and efficiencies in climate analysis and reporting, Fogarty & Machin recommended the development of automated template reporting systems.[46] Recognizing this, we developed a similar version of template reporting for the JWWP framework using the Excel programming capability. This enabled the automatic analysis of raw survey response data that was collected electronically through the Justice Agency's PPMS, and its reporting as descriptive statistics (i.e., response frequencies, percentage distributions, means and standard deviations) that were subsequently converted to indicators of risk for each climate facet. Another advantage of automation is that it enables generalist government staff to undertake the more mundane aspects of the analysis and reporting process.

JWWP outcomes are summarized and reported as tables and graphs to facilitate ease of interpretation by senior decision makers. This can take the form of heat maps of each facet and sub-facet of climate, where the 'hotter' colors indicate higher levels of risk. Table 2 provides an example of the format we used to report risk profiles for each climate facet in our pilot study of the JWWP.

High levels of employee Job Role Stress have been shown to negatively affect other aspects of climate, particularly that of employee health and well-being, psychological strain and distress, and work outcomes such as employee engagement, performance, intention to stay in the job, and therefore overall organizational effectiveness.[47] The example presented in Table 2 identifies Employee Work Overload, a sub-facet of Job role Stress, as an area of high risk for the Agency, and one requiring remedial interventions aimed at reducing the source of the risk.

Table 2: Example of JWWP climate outcome reporting – Job Role Stress

Job Role Stress Subscales	Frequency responses for Likert ratings (%)					Aggregate of 4 & 5	Response Average	Overall Risk Rating*
	1	2	3	4	5			
Role Ambiguity	50.3	34.2	7.9	4	3.6	7.6	1.7	
Role Conflict	26.7	28.9	24.4	20	0	20	2.4	
Role Insufficiency	8.9	44.3	22.2	15.5	9	24.5	2.8	
Role Overload	3.2	21.3	36.8	24.5	14.2	40	3.8	
OVERALL JOB STRESS	17.8	25.8	18.3	12.8	5.4	18.4	**2.1**	

* The risk rating is a function of the rating distributions and averages of response levels.

	RISK RATING
	High - Very High
	Moderate
	Low - Very Low

Building a Database

Psychological and organizational climate profiling of units within an agency over multiple time periods generates a large amount of data. It is important to develop a dedicated data management system to house this information for a number of reasons. In particular, it enables: (a) the benchmarking and comparison of risk profiles of units across the organization, as well as the modelling of changes over time; (b) the periodic testing, revision and continued improvement of the psychometric properties of climate measures and scales (as described above); and (c) the conduct of relational and causal modelling (discussed below) to provide deeper insight into how facets of climate impact on each other. These analyses require access to large data sets in order to provide greater analytical power and reliable outcomes.

Building the Remedial Intervention Toolkit

The logical consequence of climate profiling, particularly as suggested by the action-research model of organizational improvement, is to follow-up appropriate actions and interventions to remedy the identified problem areas. These typically involve interventions in the form of policies, strategies, initiatives or programs. In Australia and internationally there are considerable differences in the way military and government agencies respond to outcomes of employee health and well-being surveys. How and to what extent remedial actions are undertaken depend on the organizational culture, policies, *modus operandi* and HR response capability of an agency, as well as on the willingness of senior leadership to take the required actions.

With regards to the PULSE, the role of the PSS psychologist is to report outcomes to Commanding Officers in a way that will help them interpret the findings in light of the unique characteristics of their unit, and to assist them to make decisions on what actions to take– it is not their role to advise the CO about what directions they should take as a consequence of the PULSE outcomes unless requested by the CO.[48] This process reflects the predominant role of the Unit Commanding Officer, and the supportive function of PSS psychologists, in the ADF chain of command.

In contrast, senior OH&S specialists in the government sector are normally tasked to play a more proactive role in the advisory and remediation process. This may involve providing substantial recommendations as well as implementing actions aimed at managing identified risks to personnel health and well-being. Yet, as Fogarty and Machin pointed out, when it comes to addressing outcomes of large scale employee surveys there still tends to be a lack of formal connection with any guidelines or protocols for follow-up actions and interventions.[49] In order to address this shortfall, and to enable the JWWF to function as an action-research system, we developed protocols for generating recommendations for appropriate organizational level interventions aimed at treating and reducing identified climate risks.

We developed this capability for the JWWF as a remedial intervention toolkit by matching the Agency's existing HR and OH&S resources against the facets of psychological and organizational climate that are measured by the JWWP. A prototype of the toolkit is presented in Table 3.

It should be noted that most of the intervention resources listed in the toolkit are organizational level interventions that aim to affect positive changes in climate. The exceptions are individual level interventions such as Personnel Counselling, Mentoring and Performance Coaching. With these types of interventions care should be taken to apply them without identifying specific individuals at-risk, and thus breaching confidentiality assurances to survey participants. Although not ideal, this could be undertaken by more intensely promoting voluntary take-up of these services to units that have been identified at the aggregate level as being at higher risk in areas such as mental health, motivation, morale, job stress, psychological distress and strain.

Table 3: The JWWP Remedial Intervention Toolkit – Key Agency OH&S intervention initiatives and training programs matched against facets of psychological and organizational climate

Agency HR Intervention Programs	Workplace Factors										Employee Attitudes, Mental Health and Well-being					
	Job Role Stressors	Teamwork and Cohesion	Supervision	Team Communication	Organisational Support	Fairness and Autonomy	Safety Climate	Standards of Workplace Behaviour	Training Support	Confidence in Leadership	Job Satisfaction	Work Motivation	Psychological Health & Wellbeing	Workplace Energy/Fatigue	Organisational Commitment	Job and Work Engagement
Job Planning & Analysis	X	X	X	X	X	X	X	X	X	X	X	X	X	X	X	X
Effective job design	X	X	X	X	X	X	X		X	X	X	X	X	X	X	X
Task analysis	X	X	X	X	X	X	X		X	X	X	X	X	X	X	X
Workplace Health & Safety	X				X		X			X			X	X		X
OH&S policies, guidelines & regulations	X				X		X			X			X	X	X	X
OH&S management strategies	X		X		X		X			X			X	X	X	X
OH&S auditing system	X				X		X			X			X	X	X	X
Workplace risk assessments					X		X			X			X	X	X	X
Workplace safety inspections & enforcement					X		X		X				X	X	X	X
Workplace safety training (general & role specific)					X		X		X				X	X	X	X
Employee Mental Health & Wellbeing	X	X			X			X	X	X	X	X	X	X	X	X
Staff health, wellbeing & resilience training	X	X			X				X	X	X	X	X	X	X	X
Mental health awareness training	X	X			X			X	X	X	X	X	X	X		X
Stress management training	X	X			X				X	X	X	X	X	X	X	X
Employee health checks & individual health plans	X	X			X				X	X	X	X	X	X	X	X

(Continued)

Table 3: Continued

Category														
Employee counselling services	X	X						X		X	X		X	X
Leadership & Management Skills Training	X	X	X	X		X	X	X	X	X	X	X	X	X
Leadership & management training	X	X	X			X	X	X	X	X	X	X	X	X
Leadership OH&S training	X			X		X		X	X	X	X	X	X	X
Creating workplace learning cultures		X	X	X		X	X	X	X	X	X	X	X	X
Developing your staff	X	X	X	X		X	X		X	X	X		X	X
Project management skills	X	X	X	X		X			X	X	X	X		X
Professional Development &Workplace Skills Training	X	X	X	X		X	X	X	X	X	X	X	X	X
Workplace communication skills	X	X	X	X		X	X	X	X	X	X	X	X	X
Dealing with change	X	X		X		X		X	X		X	X	X	X
Indigenous & cultural awareness		X	X	X		X	X	X		X	X	X	X	X
Emotional intelligence	X	X	X	X		X	X	X	X	X	X	X		X
Workplace conflict resolution		X	X	X		X	X	X	X		X	X	X	X
Time management skills	X	X	X			X		X	X	X	X	X	X	X
Employee mentoring & performance coaching	X		X	X		X	X	X	X	X	X	X	X	X
Organisational Justice & Standards of Workplace Behaviors	X	X	X	X		X	X	X	X	X	X	X	X	X
Promoting the organisations code of conduct	X	X	X	X	X	X	X		X	X	X	X	X	X
Standards of workplace behavior & ethics training	X	X	X	X		X	X	X	X	X		X	X	X
Fraud awareness training				X	X		X	X					X	
Promoting & supporting Protected Disclosure Programs		X		X		X							X	

Employee training and development programs make up a large part of the JWWP remedial intervention toolkit because these have been shown to be effective and efficient in bringing about positive organizational change, improvements in personnel health and well-being and workplace behaviors.[50] For example, training interventions for promoting employee health and well-being include mental health awareness, employee stress management, and resilience training. Less obvious, but also important for their indirect impact on employee health and well-being because of their relevance to preventing and managing job stress, are skills training in areas such as leadership and management, communications, emotional intelligence, dealing with change, workplace conflict resolution and time management skills.

Evaluating the Effectiveness of Remedial Interventions

Benchmarking methods are useful for providing an indication of changes in climate over time and across different areas of an organization. However, they are limited in their ability to provide reliable measures of the effectiveness of specific remedial interventions. To achieve the later, we built into the Framework an optional impact evaluation capability that functions as a feedback mechanism for assessing the effectiveness of any actions taken in reducing identified problem areas, as well as for informing follow-up remedial strategies. This evaluation component is meant to contribute in a systemic way to the continuous improvement of psychological and organizational climate in line with action-research principles of organizational improvement.[51]

In the pilot trial of the JWWF, we used a basic evaluation procedure with the intention to further develop this capability in future applications. The method involved a simple repeated measures assessment of facets of climate before and three months after the implementation of remedial actions taken to improve identified areas at risk. Before and after measures were conducted close in time in order to reduce the probability that any observed changes were the result of changes in the organizational environment other than those of the remedial interventions being assessed. Although the results indicated significant improvements in the targeted areas of employee health and well-being after the interventions (indicated by lower risk scores), the low response rates and small sample size obtained for the repeated measures reduced our confidence in the results. We attributed this to the effects of survey fatigue resulting from repeating a long survey on the same employee population at points too close in time.

Given the availability of adequate resources, we recommend the application of more rigorous and evidence-based methods of evaluation for assessing the effectiveness of intervention strategies. In particular, experimental and quasi-experimental control group designs, also referred to as 'scientific' or 'evidence-based' evaluations, are the most suitable for this purpose. These methods are regarded as the gold standard of evaluation,[52] and are widely used by governments and NGOs world-wide for evaluating the effectiveness of health and well-being intervention programs.[53]

As our intervention toolkit is comprised in large part of training interventions, for our purpose Kirkpatrick's Four Level Model (FLM).[54] offers a useful set of techniques designed specifically for evaluating the effectiveness of training programs. The model consists of a framework for assessing four key levels or stages of learning impact that are important in the take up process – level one (reactions of learners) relates to participants response to the training in terms of their satisfaction with the program, including the quality of the delivery mode, learning environment and relevance; level two (knowledge gained) refers to the assessment of declarative knowledge and skills gained by participants as a result of the training; level three (on-the-job behavior change) seeks to measure the extent of transfer and application of the knowledge and skills gained to the workplace, and; level four (organizational impact) is concerned with identifying higher order organizational outcomes that result from the training, such as aggregate changes in staff health and well-being, work engagement, performance and productivity.[55]

Addressing the first two levels of the FLM can help identify specific barriers and enablers to the uptake of new knowledge.[56] The third and fourth levels of the model are useful in indicating the extent that training interventions have impacted on specific climate facets that are targeted by the training, and consequently on higher level outcomes relating to organizational effectiveness. The FLM can therefore provide a useful evaluation tool for gauging the extent to which training interventions are having the desired impact, as well as informing the adjustments that need to be made if they have been demonstrated not to be working as intended.

The evaluation capability, in combination with the profiling component and intervention toolkit, provides the feedback mechanism for the Framework to function as a research-action system for the continuous improvement of psychological and organizational climate. However, because evidence-based evaluation methods are research intensive, requiring specialist input and sufficient resourcing, the support of agency leaders to its undertaking is necessary for achieving successful outcomes.

Predictors and Outcomes of Personnel Health and Well-being: Relational Modelling as a Value-Added Capability

While psychological and organizational profiling can provide us with an indication of the level of risk for each factor, it does not tell us anything about the relationships between the domains and factors, and how they impact on each other. To obtain a deeper understanding of the dynamics of climate, 'causal' or 'relational' modelling can be undertaken. This involves the application of statistical methods from simple correlation analysis to more advanced multivariate modelling such as multiple regression, path analysis and structural equation modelling (SEM).[57]

Identifying and unpacking the key relationships between the domains and facets of climate, how they affect each other as predictors, moderators, and mediators, can add value to the Framework in several critical ways. For instance, insight derived from relational modelling can contributes to the ongoing process of revision and improvement of the conceptual model of climate by allowing us to re-examine the continued relevance of its core facets, scales, and subscales, as has been periodically undertaken with the ADF PULSE.[58]

They can also aid in informing more comprehensive intervention strategies for the smarter targeting of identified problem areas. For instance, the model presented in Figure 4, below, indicates that the psychological stress-strain relationship is not simply direct, but mediated by a number of facets such as Leadership Behaviours and Effectiveness, Unit Cohesion and Personnel Morale. This knowledge is useful for informing the engagement of broader intervention strategies through the application of the intervention toolkit (described above) to better target identified risks. For example, the model presented in Figure 4 suggests the value of promoting general leadership and management skills, in addition to those directly relevant to employee health and well-being such as stress management training, for reducing workplace stressors.

A preliminary indicator of the strength of the relationship between climate facets and domains can be provided by Correlation Analysis. Outcomes of analyses of PULSE and JWWP data consistently demonstrate moderate to strong relationships between workplace factors and those of employee health and well-being, and of the latter with work outcomes, indicating the central and critical role that personnel mental health and well-being play in psychological and organizational climate. For example, JWWP outcomes indicated strong correlations between workplace factors such as that of Job Role Stressors, and general psychological health as measured by the K10 scale ($R= -.48$, $N=98$).

Additionally, this relationship has been observed to cascade across climate domains to effect work outcomes such as Work Engagement (R=-.38, N=98).

However, the relationship between climate domains is not linear. The conceptual model we presented in Figure 2, above, while useful for climate profiling, is a simplification of the relationship between facets. Rather, climate facets interact in complex ways as predictor and outcome variables, often interchangeably, as well as functioning as psychological and organizational resources that further mediate and moderate relationships.[59] Figure 3 represents a theoretical model of the relationship between and within key domains.

Considering the many psychological and organizational climate facets at play, these types of models can be very complex. And because of common method variance in climate survey ratings, at the very least they may be used to represent models of the relationship between subjective perspectives on different aspects of the workplace. On a more ambitious level, as Albrecht et al. point out, "systemic research programs and rigorous evaluation processes are now needed to test the relationships modeled", and although they are "difficult to test as a whole, structural equation modelling can be used to test the more focused relationships embedded within the model", including the determination of "direct and indirect effects".[60]

In addition to SEM, other advanced methods of structural modelling and multivariate analytic techniques, such as multiple regression and path analysis, can help to unpack the complex relationships between facets using the same data collected for constructing the climate profiles outlined above. In structural analysis it is standard practice to begin with a simple hypothesized model that has been informed by the research literature, followed by a revised model based on outcomes from the analysis. Figure 4 provides an example of a hypothesized

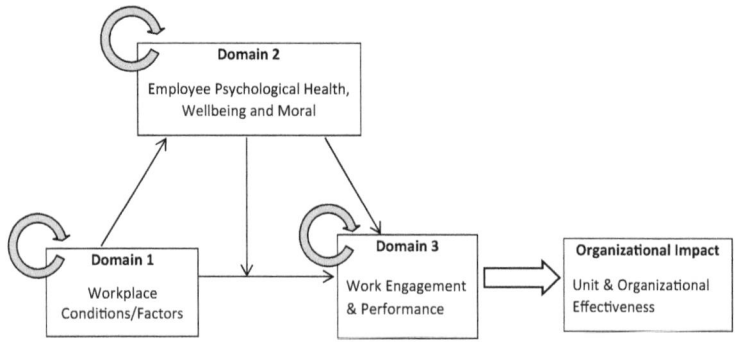

Figure 3: Theoretical model of the relationship between psychological and organizational climate domains.

Figures 4: Example of a structural model of the relationship between 'Job Stress' and 'Psychological Strain' in an ADF population, with the mediating effects of psychological resources, as derived by Murphy and Fogarty.[61]

structural model and the outcome model resulting from a path analysis of climate data from an ADF population.[62]

As already discussed, we can see in the example above that the impact of 'job stressors' on 'psychological strain' is mediated by a number of key facets of climate, such as leadership behaviors, leadership effectiveness, unit cohesion and soldier's morale, that function as organizational and psychological resources in this relationship.

Although relational modelling promises to advance the science base of psychological and organizational climate systems such as the JWWF, it is a resource and research intensive activity requiring specialist psychometric and data analytic input which most government agencies do not have the appetite or capability to undertake in-house. The value of this capability is also not well enough understood in most government organizations to attract the required resources for its undertaking. Defence Force organizations are the exception, however, as those like the Australian DoD and Canadian DND have the in-house expertise and long tradition of R&D in organizational psychology and psychometrics, particularly in areas of personnel mental health and well-being, to undertake this specialist work. Therefore, as a practical alternative to developing this capability as a standard tool for the JWWP in the non-military government sector, we propose that this function instead be considered as an optional feature to be undertaken on an as-needed basis through consultation with specialists in the field.

Summary and Conclusion

Defence Force and other government agencies have sought to develop effective tools to monitor and manage psychological and organizational climate in order to promote healthier and more effective employees, teams and organizations. Such tools also enable senior leaders to demonstrate that they are implementing initiatives in accordance with evolving workplace health and safety legislation and regulations. In this chapter, we discussed our contribution to these efforts by developing and trialing the JWWF, a climate surveillance and intervention framework, for a large Australian government agency.

In order to address the identified shortcomings of current climate profiling systems used in the government sector,[63] we aimed to make the Framework evidence-based, user-friendly, and able to inform the implementation of actions necessary to affect continuous positive organizational change. We achieved this by imbuing the JWWF with several enabling features and capabilities.

First, we sought to apply principles of organizational psychology and psychometrics to strengthen the science-base, validity, and reliability of the JWWP, which is the profiling component of the JWWF. This process was facilitated by knowledge transfer from well-established and robust Defence Force climate profiling systems – the ADF PULSE and the CF UMP. In particular, lessons learnt from the PULSE model guided the development of a number of key aspects of the JWWP, namely: (a) the conceptual model and its domain and facet structure; (b) the construction of core measures and scales – with modifications tailored to our Agency's requirements; (c) analytical protocols for the continuous improvement of measures based on established PULSE and CF UMP psychometric practice; and (d) an optional relational modelling capability that can be used to generate knowledge and insight into how psychological and organizational climate facets interact and impact on each other.

The leap from the ADF PULSE to the JWWP was not a difficult one, as the Justice Agency for which the latter was developed is a large and diverse government organization comprising many divisions and units responsible for policy making, regulation and operational functions such as law enforcement, security and emergency services, which share similar workplace concerns to those of the Department of Defence. As such, the initiative is a good example of successful R&D knowledge sharing between military and civilian government sector agencies.

Second, we introduce additional components to the JWWF that were driven by the specific OH&S requirements of the Agency, as well as the need to expand the scope and capability of the Framework so that it could function as an action-research system for the continuous improvement of psychological and organizational climate. We achieved this by developing: (a) an automated survey delivery and data collection system connected to the Agencies existing employee performance management system, which enabled it to operate with minimal use of Agency time and resources; (b) an Agency specific remedial intervention toolkit for informing actions to address identified climate problem areas, and; (c) an evidence-based evaluation capability to assess the effectiveness of interventions in reducing identified risks, and to inform further remedial actions, if needed. Although, due to limited resourcing, our initial trial of the JWWF evaluation function was limited to applying a simple repeated-measures method, in this paper we have recommended the use of more robust methods for use in future applications.

Overall, the JWWF was successfully trialed and demonstrated to be an effective and efficient tool for the surveillance, profiling, and management of psychological and organizational climate risk in a large government agency. During this process, lessons have also been learnt that are important for ensuring the success of this and similar frameworks in future applications. These include the need to: (a) implement more effective strategies for achieving good survey participation rates of government sector employees in order to increase the analytical power and reliability of outcomes; (b) obtain the commitment and support of senior agency leaders to secure sufficient resources to run all phases of the Framework, including the evaluation component that enables it to function as an action-research tool for continuous climate improvement; and (c) following the example set by the ADF PULSE and CF UMP, establish robust protocols for the ongoing improvement of climate profiling measures. Finally, the testing of the Framework in different public sector environments is necessary to demonstrate its versatility, relevance, and usability in diverse government settings.

Notes

1 Jennifer Z. Carr, Aaron M. Schmidt, J. Kevin Ford, and Richard, P. DeShon, "Climate perceptions matter: A meta-analytic path analysis relating molar climate, cognitive and affective states, and individual level work outcomes." *Journal of Applied Psychology*, 88, no. 4 (2003): 605; Christopher Parker, Boris Baltes, Scott A. Young, and Joseph W. Huff, "Relationship between psychological climate perceptions and work outcomes: A meta-analytic review." *Journal of Organizational Behavior*, 24, no. 4 (2003), 389–416.
2 These included: Samuel A. Stouffer, Edward A. Suchman, Leland C. DeVinney, Shirley A. Star, and Robin M. Williams, Jr. Foreword by Frederick Osborn. *The American soldier: Adjustment during army life (Studies in Social Psychology in World War II)*, Vol. 1 (1949). Princeton, NJ: Princeton University Press; Samuel A. Stouffer, Arthur A. Lumsdaine, Marion Harper Lumsdaine, Robin M. Williams, M. Brewster Smith, Irving L. Janis, Shirley A. Star, and Leonard, S. Cotrell, Jr. *The American Soldier: Combat during Aftermath, Volume 2*, (Princeton, NJ: Princeton University Press, 1949). For a historical review of Stouffer's work see Joseph W. Ryan, "Samuel Stouffer and The American Soldier." *Journal of Historical Biography*, 7 (Spring 2010): 100–137.
3 A history of advancements and applications of the methods and tools of social and behavioural science in the military can be found in D.R. Segal, "Spotlight on History." *The Military Psychologist*, 32, no. 1 (2017): 20–24; For an outline of more recent developments in psychological and organizational climate tools and the Australian Defence Force PULSE, see Anne Goyne, Robert Lake, Maree Riley, and Brian Johnston, "Taking the pulse of your unit: A command support tool for assessing unit climate." In Peter Murphy (Ed.), *Focus on human performance in land operations* (2009). Canberra, ACT: Australian Government Department of Defence.
4 Anne Goyne, Robert Lake, Maree Riley, and Brian Johnston, "Taking the Pulse of your Unit: A Command Support Tool for Assessing Unit Climate".
5 K.J. Brown, M.E. Norris, and B.F. Johnston. *Unit morale profile: A psychometric analysis*, Sponsor Research Report 2005–16 (Jul 2005). Ottawa, Ontario: Director Human Resources Research and Evaluation/National Defence Headquarters/DND; D.E. Woycheshin. *Review of the unit morale profile core measures*, Technical Note 2007–11 (Jun 2007). Ottawa, Ontario: Director Personnel Applied Research/National Defence Headquarters/DND; D.E. Woycheshin. *Review of the unit morale profile toolbox measures*, Technical Note 2007–12 Jun 2007). Ottawa, Ontario: Director Personnel Applied Research/National Defence Headquarters/DND; Maxime A. Tremblay. *Unit morale profile: A psychometric analysis*, DGM TM 2009–013 Sept (2009). Ottawa, Ontario: Defence R&D Canada/Director General Military Personnel Research & Analysis/Chief Military Personnel, DND.

6 Gerard Fogarty. *Psychometric review of the PULSE survey* (Version 4404), Technical Brief 25/2012 (Dec 2012). Canberra, ACT: DOPHA MHPRN/JHC/ Department of Defence.
7 A Price Waterhouse Coopers & Medibank Health Solutions. *Workplace wellness in Australia: Optimising the benefits of workplace wellness* (September 2010). Price Waterhouse; Beyond Blue & Price Waterhouse Coopers. *Creating a mentally healthy workplace: Return on investment analysis final report* (May 2014). Price Waterhouse.
8 State of Victoria. *The Occupational Health and Safety Act* (2004). Melbourne, Australia: State Government of Victoria, 2004.
9 Safework Australia. *Work-related psychological health and safety: A systematic approach to meeting your duties – National guidance material* (January 2019). Australian Government.
10 Jan-Louise Godfrey, "APS psychologically healthy workplace program: Helping organisations work better." *InPsych: The Bulletin of the Australian Psychological Society*, 6(36) (December 2014).
11 Paul M. Muchinsky, *Psychology applied to work: An introduction to industrial and organizational psychology*, 8th ed. (2006). Belmont, CA: Thomson Wadsworth.
12 Simon L. Albrecht, Arnold B. Bakker, Jamie, A. Grunman, William H. Macey, and Alan, M. Saks, "Employee Engagement, Human Resource Management Practices and Competitive Advantage." *Journal of Organizational Effectiveness*, 2, no. 1 (2015), 7–35; Jennifer Z. Carr, Aaron M. Schmidt, J. Kevin Ford, and Richard, P. DeShon, "Climate perceptions matter: A meta-analytic path analysis relating molar climate, cognitive and affective states, and individual level work outcomes."; Christopher Parker, Boris Baltes, Scott A. Young, and Joseph W. Huff, "Relationship between psychological climate perceptions and work outcomes: A meta-analytic review".
13 For example, SafeWork Australia. *Work-related psychological health and safety: A systemic approach to meeting your duties – additional guidance material* (January 2019). Canberra: Australian Government.
14 APSC. *State of the service report 2017–2019 – APS employment census* (2018). Canberra: Australian Public Service Commission (apsc.gov.au); Australia. VPSC. *People matters survey* (2018). Melbourne: Victorian Public Sector Commission (vpsc.vic.gov.au).
15 Gerard Fogarty and Anthony Machin, "Combining database technologies in support of a safety climate improvement strategy," *Proceedings of the 49th IMTA Conference*, Gold Coast, Australia: International Military Testing Association (October 2007), 162–170.
16 Ibid.

17 Walter L. Giusti and Gayle Sherwell. *Justice working well profile* (Report No.1). Melbourne, Australia: People and Culture Division. State Government of Victoria, April 2013.
18 Ibid.
19 Clem Adelman, "Kurt Lewin and the origins of action research." *Educational Action Research*, 1, no. 1 (1993); B. Dick, "Action research: Action and research." In S. Sankaran, B. Dick, R. Passfield, and P. Swepson, (Eds.), *Effective change management using action learning and action research* (2001). Lismore, NSW: Southern Cross University Press.
20 Walter L. Giusti and Gayle Sherwell, *Justice working well profile*.
21 For a comprehensive treatment of this issue see: Edgar H. Schein, "Sense and nonsense about culture and climate," Commentary in *Handbook of organizational culture and climate* (September 1999). Boston, MA: MIT Sloan School of Management; Cheri Ostroff, Angelo J. Kinicki, and Rabiah S. Muhammad, "Organisational culture and climate." In Irving B. Weiner, (Ed.), *Handbook of psychology*, 2nd ed. (2012). Wiley.
22 Edgar H. Schein. *"Sense and nonsense about culture and climate"*.
23 PRTG, *PULSE User's Guide*.
24 Christopher Parker, Boris Baltes, Scott A. Young, and Joseph W. Huff, *"Relationship between psychological climate perceptions and work outcomes: A meta-analytic review."*
25 Laurel L. Hourani, Thomas V. Williams and Amii M. Kress, "Stress, mental health, and job performance among active duty military personnel: Findings from the 2002 Department of Defence Health-Related Behaviors Survey." *Military Medicine*, 171, no. 9 (September 2006): 849–856; Stephen E. Pflanz and Alan D. Ogle, "Job Stress, Depression, Work Performance, and Perceptions of Supervisors in Military Personnel." *Military Medicine*, 171, no. 9 (2006): 869–865.
26 Gerard Fogarty and Anthony Machin, *"Combining database technologies in support of a safety climate improvement strategy"*.
27 Earl R. Babbie, *The practice of social research*, 5th ed. (1989). Belmont, CA: Wadsworth Publishing Company; G.W. Heiman, *Research methods in psychology* (1995). Boston, MA: Houghton Mifflin Company; J.J. Shaughnessy and E.B. Zechmeister, *Research methods in psychology*, 4th ed. (1997). Singapore: McGraw-Hill International Editions.
28 Anne Anastasi and Susana Urbina, *Psychological testing*, 7th ed. (1997). Upper Saddle River, NJ: Prentice-Hall International; Paul Kline, *The new psychometrics: Science psychology and measurement* (1998). New York; NY: Routledge; Kevin R. Murphy, and Charles O. Davidshofer. *Psychological testing: Principles and applications*, 6th ed. (2005). Upper Saddle River; NJ: Pearson Education International.

29 Walter L. Giusti, *Measurement of military post-operational fatigue: A short version of the multidimensional fatigue inventory for ADF applications* (Internal working paper prepared for the Australian Defence Force Decompression Program Evaluation Working Group). Canberra, Australia: Directorate of Strategy and Operational Mental Health, MHPRB, Australian Department of Defence, 2011.
30 Ellen M.A. Smets, B. Garsen, B. de Bronke, and J.C.J.M. de Haes, "The Multidimensional Fatigue Inventory (MFI) psychometric qualities of an instrument to access fatigue." *Journal of Psychosomatic Research*, 39, no. 3 (1995): 315–325.
31 Wilmar B. Schaufeli, Arnold B. Bakker, and Marisa Salanova, "The measurement of work engagement with a short questionnaire: A cross-national study." *Educational and Psychological Measurement*, 66, no. 4 (August 2006): 701–716.
32 Ibid.
33 M.S. Christian, A.S. Garza and J.E. Slaughter, "Work engagement: A quantitative review and test of its relations with task and contextual performance." *Personnel Psychology*, 64 (2011): 89–136; E. Demerouti and R. Cropanzano, "From thought to action: employee work engagement and job performance." In A.B. Bakker, and M.P. Leiter. (Eds.), *Work engagement: A handbook of essential theory and research*, pp. 834–848 (2010). New York, NY: Psychology Press.
34 Wilmar B. Schaufeli, Arnold B. Bakker, and Marisa Salanova, *"The measurement of work engagement with a short questionnaire: A cross-national study"*.
35 Kessler, G. Andrews, L. Colpe, E. Hiripi, D. Mroczec, S. Normand, E. Walters and A. Zaslavski, "Short screening scales to monitor population prevalence's and trends in non-specific psychological distress." *Psychological Medicine*, 32 (2002): 959–976.
36 Walter L. Giusti, *Measurement of military post-operational fatigue: A short version of the multidimensional fatigue inventory for ADF applications*.
37 Ellen M.A. Smets, B. Garsen, B. de Bronke, and J.C.J.M. de Haes, *"The Multidimensional Fatigue Inventory (MFI) psychometric qualities of an instrument to access fatigue"*.
38 For the CF UMP – K.J. Brown, M.E. Norris, and B. F. Johnston. *Unit morale profile: A psychometric analysis*; Woycheshin, *Review of the unit morale profile core measures*; Tremblay. *Unit morale profile: A psychometric analysis*. And, for ADF PULSE – Gerard Fogarty. *Psychometric review of the PULSE survey*.
39 Walter L. Giusti and Gayle Sherwell, *Justice working well profile*.
40 Gerard Fogarty, *Psychometric review of the PULSE survey* (Version 44004).
41 Maxime A. Tremblay, *Unit morale profile: A psychometric analysis*.
42 Joseph F. Hair, William C. Black, Barry J. Babin, and Rolph E. Anderson. *Multivariate data analysis: A global perspective*, 7th ed., pp. 92 & 125 (2010), Upper Saddle River, NJ: Pearson Prentice Hall; Ronald J.Cohen and Mark

E. Swerdlik. *Psychological testing and assessments: An introduction to tests and measurement*, 7th ed., p. 151 (2010). Boston, MA: McGraw-Hi; Murphy and Davidshofer. *Psychological testing: Principles and applications*, pp. 133-134 & 142.
43 Tremblay (Sept, 2009), Fogarty (Dec, 2012), and Giusti and Sherwell (2013), respectively.
44 APSC, *State of the service report 2017-2019 – APS employment census*; VPSC *People matters survey*.
45 Cohen, J, "A power primer." *Psychological Bulletin*, 112, no. 1 (1992): 155–159.
46 Gerard Fogarty and Anthony Machin, *"Combining database technologies in support of a safety climate improvement strategy"*.
47 Laurel L. Hourani, Thomas V. Williams, and Amii M. Kress, *"Stress, mental health, and job performance among active duty military personnel: Findings from the 2002 Department of Defence Health-Related Behaviors Survey"*.
48 PRTG. *PULSE user's guide* (2006). Canberra, ACT: Psychology Research and Technology Group, Defence Force Psychology Organisation.
49 Gerard Fogarty and Anthony Machin, *"Combining database technologies in support of a safety climate improvement strategy"*.
50 Herman Aguinis and Kurt Kraiger, "Benefits of training and development for individuals and teams, organizations and society." *Annual Review of Psychology*, 60 (2009): 451–474; Arthur Winfred Jr., Winston Bennett Jr., Pamela S. Edens, and Suzanne T. Bell, "Effectiveness of training in organizations: A meta-analysis of design and evaluation features." *Journal of Applied Psychology*, 88, no. 2 (2003): 234–245.
51 Sankaran, S., Dick, B., Passfield, R. and Swepson, P. (Eds.), *Effective change management using action learning and action research* (2001). Lismore, NSW: Southern Cross University Press.
52 Peter H. Rossi, Mark W. Lipsey, and Howard E. Freeman. *Evaluation: A systemic approach* (2004). Thousand Oaks, CA: Sage Publications; Earl R. Babbie, *The practice of social research*; J.J. Shaughnessy and E.B. Zechmeister, *Research methods in psychology*.
53 US Government Accountability Office. *Quantitative data analysis: An introduction – report to program evaluation and methodology division* (GAO/PEMD-10.1.11, May 1992); US Government Accountability Office, *Applied research and methods: Designing evaluation* (GAO-12-208G, Jan 2012).
54 D. Kirkpatrick. Evaluating Training Programs: The Four Levels, 2nd ed. (San Francisco; CA: Berrett-Koehler, 1998); Tamkin J. Yarnall, and M. Kerry. Kirkpatrick and Beyond: A Review of Models of Training Evaluation (Brighton: Institute of Employment Studies, 2002).
55 H. Steensma and K. Groeneveld, "Evaluating training using the four level model." *Journal of Workplace Learning*, 22, no. 5 (2010): 319–331; E. Salas and

J.A. Cannon-Bowers, "The science of training: A decade of progress." *Annual Review of Psychology*, 52, no. 1 (2001): 471–499.
56 B.D. Blume, J.K. Ford, T.T. Baldwin, and J.L. Huang, "Transfer of training: A meta-analytic review." *Journal of Management*, 36, no. 4 (2010): 1065–1105; L.A. Burke and H.M. Hutchins, "Training transfer: An integrative literature review." *Human Resource Development Review*, 6, no. 3 (2007): 263–296.
57 Rex B. Kline. *Principles and practice of structural equation modeling*, 4th ed. (2016). New York, NY: The Guilford Press. For more general texts of these methods see: Joseph F. Hair, William C. Black, Barry J. Babin, and Rolph E. Anderson. *Multivariate data analysis: A global perspective* (2010); Barbara G. Tabachnick and Linda S. Fidell. *Using multivariate statistics*, 5th ed. (2007). Boston, MA: Pearson Education.
58 Gerard Fogarty, *Psychometric review of the PULSE survey (Version 44004)*.
59 Simon L. Albrecht, Arnold B. Bakker, Jamie, A. Grunman, William H. Macey, and Alan, M. Saks, *"Employee engagement, human resource management practices and competitive advantage"*; Jennifer Z. Carr, Aaron M. Schmidt, J. Kevin Ford, and Richard, P. DeShon, *"Climate perceptions matter: A meta-analytic path analysis relating molar climate, cognitive and affective states, and individual level work outcomes"*; Christopher Parker, Boris Baltes, Scott A. Young, and Joseph W. Huff, *"Relationship between psychological climate perceptions and work outcomes: A meta-analytic review"*.
60 Simon L. Albrecht, Arnold B. Bakker, Jamie, A. Grunman, William H. Macey, and Alan, M. Saks, *"Employee engagement, human resource management practices and competitive advantage"*, 21.
61 P. Murphy and Gerard Fogarty, "Leadership: The key to meaning and resilience on deployment." In Peter Murphy (Ed.), *Focus on human performance in land operations*, pp. 95–104 (2009). Canberra, ACT: Australian Government Department of Defence.
62 Ibid.
63 Gerard Fogarty and Anthony Machin, *"Combining database technologies in support of a safety climate improvement strategy"*.

Petrus C. Bester

A Positive Psychology Perspective on Predeployment Fitness-For-Duty Evaluations for External Deployments: A Proposition for the South African National Defence Force[1]

Introduction

As a member state of the United Nations (UN), the African Union and the Southern African Development Community (SADC), South Africa is in the vanguard of bringing about lasting peace and stability on the continent.[2] In recent years the South African National Defence Force (SANDF) has participated in peace-support operations in various African states such as Burundi, Sudan and the Democratic Republic of the Congo. Consequently, South African soldiers deploy in the African battlespace (ABS), which can be described as an environment where soldiers are increasingly placed in asymmetric situations against opponents who are not easily identifiable and who are probably better armed and equipped, with better access to communications and technology.[3] Various authors have reported on the changed nature of the ABS and confirmed the complexity and challenges resulting from these changes.[4]

These are complex situations and Gouws[5] emphasizes that the complexities of military operations domestically and abroad require soldiers who are more resillient than ordinary citizens. Moreover, soldiers are very likely to be exposed to psychological trauma that inflicts a shock to the central nervous system, resulting in a reduction in the connectivity between limbic and cortical processes.[6] The individual's ability to recover from such trauma depends on his or her previous history, physiological status and the social distress encountered.[7] Furthermore, the demands of the SANDF to deploy into peacekeeping and operations other than war scenarios to satisfy UN and SADC needs on the African continent lead to an increased operational tempo for soldiers, thus a greater likelihood of exposure to trauma, as well as prolonged deployment periods from home that may exceed the SANDF's typical deployment duration of twelve months for peace support operations.

For these reasons the UN has also indicated that deploying soldiers must undergo adequate medical, dental and/or psychological screening prior to deployment, implying that they should have "no psychopathology".[8] It is therefore understandable that Cabinet decided in 1998 that South African soldiers must undergo once every two years health assessments that include the psychological state of soldiers[9] to ensure that they are combat-ready. That decision compelled the SANDF's Directorate Psychology to screen soldiers prior to deployment.

After an initial review of the literature, the SANDF's Director Psychology decided to screen for general psychopathology in the pre-deployment phase.[10] This is consistent with international trends to screen for pathology.[11] In the absence of a South African instrument or tool for assessing pathology the Millon Clinical Multiaxial Inventory III was used initially. However, because of high cost and low scale reliabilities when applied to a South African population, the instrument was later rejected. A decision was made to "screen out", which implied the identification of mental health issues that could adversely affect performance as opposed to "screening in", when candidates that meet specific criteria are sought.[12] This implied screening for the "three Ds", being disease, disorder and disability.[13] The Military Psychological Institute (MPI) subsequently developed a psychopathology screening instrument, but the matter was further complicated by a situation where a supposedly "normal population" or "non-psychiatric population" whose members are most likely highly functional in the workplace, with no known diagnosed psychological disorders, had to be screened.[14] In this context "normal" refers to a "non-psychiatric" population consisting of individuals who are highly functional in the workplace with no known diagnosed psychiatric disorders. Moreover, the absence of psychopathology does not imply the presence of mental health and *vice versa*.[15]

This chapter consequently embraces a positive psychological approach, rather than relying on a restrictive medicalized framework that would tend to only seek pathology. Utilizing a purely medical model in the study of human experiences may lead one to neglect other psychological risk factors that may have a negative impact on psychological well-being (PWB) that can be identified and addressed by interventions other than psychotherapy. After all, it needs to be acknowledged that there are conditions other than medical disorders that will make a person a psychological risk for deployment.

Consequently, the author will provide background on the current screening process, which places stronger focus on screening for psychopathology. This chapter will also compare psychopathology with psychological risk, discuss positive psychology, propose an integrative military model for soldiers' well-being in the external deployment context as basis for a future process of screening for deployment, and conclude the discussion with recommendations for decision-makers in the SANDF.

The Current Screening Process

The screening process in the SANDF is referred to as the "Concurrent Health Assessment"[16] (CHA), whereas the broader military psychology literature refers to it as "military-fitness-for duty-evaluations".[17] References to the CHA, are closely linked to fitness-for-duty evaluations. The aim of these evaluations is to select those soldiers whose fitness for military duty would be preserved when operationally active, even in some extreme situations. From the perspective of military psychology, it means that they can perform their jobs safely and effectively from a mental health or neuropsychological perspective. It is even possible for a soldier to be medically classified without a so-called health board owing to the process followed during the assessment.[18] Moreover, it is important to note that the CHA does not include post-deployment assessments, forensic assessments, security vetting assessments and medical evaluation boards. As stated above, its focus is routine assessments to determine operational deployability.

The current process requires psychological and social work assessment of the soldier to be done preferably one week before the medical assessments are to begin, although it is not always possible in practice. When the remainder of the process starts, the soldier's blood is drawn, an oral health assessment is done by a dentist, as well as medical screening including HIV/AIDS pre-test counselling, pre-medical assessments to obtain clinical data, and other physical medical assessments. These procedures also include the immunization of the soldier or confirmation that immunizations are up to date and will remain up to date for the duration of the anticipated deployment period.

During the psychological assessment soldiers complete the Psychological Risk Inventory (PRI)[19] that was developed by the MPI as a screening tool to assess the psychological health status of SANDF members when a clinical psychologist is required to assess the mental health status of large groups of people within a short period of time. It is important to note that the PRI is not a diagnostic tool, but a self-report measure that attempts to identify behavioral patterns that differ significantly from a so-called normal SANDF population. To conduct a clinical interview with each individual is very time-consuming and the inventory is thus used as a cost-effective tool to identify risk factors.

According to the MPI the PRI measures a number of primary, secondary and latent scales to assess the overall risk behavior of soldiers in a normal population.[20] The primary scales measure whether the respondents (the soldiers in this instance) are under *stress*; whether they are suffering as a consequence of past *trauma*; and whether they display *aggression,* (by responding in the affirmative to questions with an aggressive and/or impulsive content, for example, being irritated by others). The secondary scales refer to *withdrawal,* which represents a mixture of withdrawal, anxiety and depression indicators; *mood,* which refers to behavioral indicators associated with negative moods; *dissociative* traits, which are indicative of absent-mindedness, possible depression and inability to cope with day-to-day stressors; and *destructive behavior,* which is a composite scale with diverse indicators of an ineffective coping strategy. Lastly, the latent scales consist of: *mild anxiety* indicative of constant worry and feelings of being down-hearted and/or irritated for no reason; *low ego integrity,* which is indicative of a person's ego being under pressure based on vague and non-specific complaints; and lastly, *general negative behavior,* which is not a scale on its own, but contains items that can provide valuable information that the psychologist can use if the respondent is identified for an interview.

In addition to these scales the questionnaire includes a *risk indicator* consisting of a combination of all the items related to the primary, secondary and latent scales, but excluding items relating to moderating indices. There is also a *coping indicator,* which is a moderating value that allows the psychologist to make an informed decision on the state of the respondent's coping mechanisms in relation to the risk factors reported in the test.

Those with identified risk factors are then interviewed by a psychologist and a determination is made with respect to their health status in relation to deployability. Based on a statistical formula, about 23 % of those assessed will be identified for a clinical interview.[21] The psychologist then makes a determination with respect to the mental health status of the individual in relation to external operational deployability according to a color code assigned to each individual.[22] These codes are "green", "yellow" and "red" and indicate the soldier's external deployability.

A green operational status indicates no restrictions on the utilization of the individual based on the confirmed medical (health) category. Yellow status implies a temporary restriction on deployment. In this case an intervention may enable the status to be changed to green after conditions change as a result of the intervention. For example, an obese person who is assessed as yellow status goes on a diet, loses weight and is then determined to be within the norms set by the SANDF. When a soldier has a permanent inability (irreversible condition) to deploy operationally externally, as in the case of kidney failure, red status will be attributed to him or her. During internal deployments, decisions will be based on a case-by-case assessment in relation to the nature of deployment, occupational class and the individual health profile of the soldier.

Essentially, the color-coded system reflects a representation of the soldier's health status and is used to initiate follow-up or interventions in respect of further treatment. As alluded to above, this color coding is done not only holistically, but also per discipline or component of the process. A person may be "green" in terms of his or her physical health, but may be "yellow" because of a mental health related issue, which then will holistically make the person unsuitable for deployment and an intervention would be required to change the person's status (color code). All disciplines involved, such as psychology, social work, medicine and oral health, capture the interpreted results electronically and these are imported back on the South African Military Health Service's (SAMHS) information technology system where they are accessible for a final decision to be made on a member's suitability for deployment.

Consequently, on completion of the whole process, a so-called confirming health care practitioner (HCP) evaluates all available information (inputs from all the disciplines involved) and determines the health status of the member. This HCP may require that additional medical information be obtained if an informed decision is not possible. Furthermore, all the information is captured on the SANDF's Health Informatics System for future reference and statutory record-keeping requirements. Discipline-specific information is only available to that discipline's practitioners who are registered on the system and to the confirming HCP. Patient information is not accessible across different disciplines, for the sake of confidentiality.

The author's criticism against the existing process is that it is mainly restricted to medical information and no other collateral information is considered for determining deployability, for example what is manifested in a person's behavior on a day-to-day basis. Consequently, another approach from a positive psychology perspective is suggested. However, before conceptualizing the alternative approach, it is important to contrast the concepts of psychopathology vs psychological risk.

Psychopathology Versus Psychological Risk

Psychopathology resides within the field of abnormal psychology and refers to the study of the symptoms and causes of mental distress and various treatments for behavioral and mental disorders.[23] Focusing on psychopathology in relation to deployment may lead to the medicalization of normal behavior.[24] Therefore, another approach is required, and the concept of psychological risk seems attractive, as it relates better to a mental health model than to a medical or disease model. Furthermore, a person who is a psychological risk is not necessarily ill in terms of a mental health perspective where psychopathology is applicable. An overview of the psychology and risk-related literature provided a limited description of the concept of psychological risk as it might be applicable to military deployments.

From the perspective of occupational health and safety, psychological risk is seen as emanating from the work environment, namely poor work design or poor social context of the work, leading to physical and social outcomes such as work-related stress, burnout or depression.[25] Born and van der Flier[26] identified individual differences when they investigated organizational risk and observed that there are relevant aspects related to individual differences such as coping styles (how people cope with the events they encounter). From a national security perspective, Wiese[27] refers to the concept of security risk, which suggests that because of a person's personality traits, needs, behavior, ideological persuasion or extreme sensitivity in terms of past deeds, he/she can be persuaded by whatever means to cooperate with an unauthorized individual or organization to divulge secrets of his/her employer or of his/her own accord would divulge secrets to an unauthorized individual or organization. Based on the work of Wiese, the author of this chapter, who also does consulting work related to compiling a risk profile of existing or prospective employees, conceptualized the "risk profile" of an individual as the outcome of a process to determine whether an individual poses a risk to his or her employer or prospective employer. This is based on an analysis of his or her personality traits, life style, needs, behavior, value system or sensitivity in terms of past deeds and whether he or she can be persuaded by whatever means to cooperate with an unauthorized individual or organization to cause intentional or unintentional damage to his or her company or any of the company's clients.[28] This definition can be viewed as "employment risk".

McSherry[29] discusses psychological risk from a mental health perspective in a criminology setting and states that it is about identifying the risk of dispositions that may lead to reoffending or harming others. Reference is made to a combination of both a clinical and actuarial approach to decision-making on psychological risk and a number of predictor variables have been identified from the literature. These variables are past violence (relying on the history of past behavior); pre-existing vulnerabilities such as early signs of antisocial traits, difficulties in peer relationships and hostility towards authority figures; social and interpersonal factors; mental illness; substance abuse, especially when it co-exists with mental illness; state of mind; situational triggers such as loss, demands, expectations, confrontations, ready availability of weapons, and physical illness; and personality constructs. One can also see the relationship between these pre-existing vulnerabilities and the content of the scales of the PRI discussed above. Moreover, in an effort to deal with a normal population, the MPI states that psychological risk "... *refers to the behavioral indicators of [an] ineffective coping mechanism to deal with the psychological demands of everyday life.*"[30] The notion is that if a person does not demonstrate the necessary coping strategies, the likelihood of the development and progression of a mental health disorder is increased if the problem is not identified and treated timeously. Taylor's description of coping strategies, as cited by the MPI, refers to the specific efforts, both behavioral and psychological, that people employ to master, tolerate, reduce, or minimize stressful events.[31] There are a number of generalized and specific resources that can assist an individual when attempting to cope with various stressors.[32] These include self-respect, cultural values, tradition, intelligence, view of the world and healthy behavior. According to Antonovsky,[33] these resources can be classified into eight groups, namely physical, biochemical, material, cognitive, emotional, estimation and views, interpersonal relational, macrosocial and cultural.

In the military context the stress caused by stressful events is defined as "... *a non-specific number of reactions and responses of the body (mental, emotional and/or physiological) causing bodily or mental tension, strain or pressure following any demand made upon it, and designed for self-preservation.*"[34] Stress can be either good (referred to as eustress) or bad (referred to as distress) and can be caused by internal or external forces.[35]

When analyzing the aforementioned definitions related to psychological risk, a number of broad themes come to the fore, namely context (social, work, personal), predictive variables, coping, and outcomes. Considering these themes and for the purpose of this discussion, psychological risk can be defined as:

> ... *the sum total of the outcomes emanating from an individual's ability to cope with contextual encounters as determined by predictive variables affecting that specific individual.*

The context refers to, but is not limited to, the social, work and personal environment of the soldier, while the predictive variables include, but are not limited to, that individual's past behavior, pre-existing vulnerabilities, mental illness, social and interpersonal factors, substance abuse, situational triggers, and certain personality constructs. Should a soldier deploy to a mission area in for example Sudan, the context would be the foreign country, Sudan, with all its characteristics, such as climate, culture, population and social system, his or her personal circumstances at home and his or her experience of working as soldier as part of a team interacting with the social system in Sudan.

With a richer understanding of what a psychological risk is and knowing that it is directly related to the outcomes of an individual's ability to cope, one can then ask what the consequences or outcomes of psychological risks are, especially those that lead to unsuccessful deployment, which is in itself also a negative outcome or consequence. In essence, when a soldier becomes a challenge for his or her commanders during deployment, for example by being charged through the legal system and/or having to be prematurely repatriated because of health or behavior-related issues, the deployment for the member is viewed as unsuccessful. From the above, one can infer that there are numerous psychological risk factors that are likely to contribute to unsuccessful deployment by an individual or can at least increase the risk of having an unsuccessful deployment if the person cannot cope with it. Such unsuccessful outcomes may stem from psychopathology such as suboptimal intelligence, anxiety disorders, adjustment disorders (lack of adaptability), lack of motivation, a history of ineffectiveness, difficulties with interpersonal relationships, failure to seek mental health assistance when needed, or inappropriate personal conduct – including criminal behavior, financial difficulties, incidents involving substance abuse, poor performance, and even incompatibility with the military culture.

Thus, negative outcomes may not stem from psychopathology only, because psychological risk also includes other dysfunctional behaviors that cannot be diagnosed as psychopathology. Some of the typical dysfunctional work behavior is absenteeism, substance abuse, presenteeism (functionally absent although physically at work), theft, bullying and sexual harassment.[36] Unmanaged psychological risks may also lead to inability to manage the demands of the job[37], the development of mental health problems (psychopathology), increased safety-related incidents such as psychological and sexual harassment and third party violence[38], uncontrolled individual behavior that may lead to an international incident and ultimately diplomatic embarrassment during for example a peace-support operation. It can also have a negative impact on unit cohesion that restricts a military force's ability to function optimally. From a private business perspective, one would state that the outcome would be overall poor business performance, whereas in the military it would most probably denote a failed mission.

Based on the preceding discussion one can postulate how the changed ABS may place new demands on South African soldiers. In support, Rothmann, Mostert and Strydom[39] emphasize that work has an impact on the well-being of employees and that, depending on the unique resources and demands in a specific work context, the determinants of well-being may differ in various working environments. Moreover, every occupation has its own specific risk factors regarding well-being and the military is one of those places where the risks are higher and more. This confirms the conception that deployment is emotionally demanding, and that emotional coping is required. Happy, resilient and coping soldiers are likely to function optimally.

From the above it is clear that despite the advancements made in the pre-deployment assessments in terms of moving from screening for psychopathology only to including psychological risk, the process is in essence still one focusing on maladaptation in its many and varied forms. It is therefore necessary also to focus on and include other risk factors in the psychopathology-free majority of the population of soldiers and assist them to enjoy life more, become more productive, and develop a sense of positive engagement and meaning during deployments specifically and in life in general. Therefore, the next section will provide more insight into the positive psychology approach.

A Positive Psychology Approach

From the discussion above, an attractive option is to approach pre-deployment screening from a positive psychology framework. Positive psychology is the application of psychological principles to real-life issues through attempting to understand the causes and consequences of optimal human functioning to help one manage and succeed in the workplace, deliver better and more compassionate health care, and provide effective and engaging education in ways that optimize achievement (psychological well-being, and the development of community).[40] Mathews[41] describes positive psychology as a paradigm that provides a systematic conceptual basis and groundwork for an empirical assessment of the role of character strengths in military adaptation and performance. It is thus about efficacious adaptation and excellence in all domains of life and therefore has utility within the domain of military psychology where military commanders would like to command high-performing soldiers. It is for this reason that Milnič[42] investigated the application of the salutogenetic paradigm[43], which also falls within the field of positive psychology that focuses on health and well-being in the field of military psychology. Others, such as Bartone, also studied hardiness, another construct related to positive psychology that refers to a collection of personality characteristics that functions as a flexible resource during encounters with demanding life events.[44]

Positive psychology is not meant to replace traditional methods and models employed by military psychologists; it rather entails adding concepts and methods to the military psychologist's toolbox.[45] The functionality of positive psychology in the military is further supported by Matthews[46] when he states that the military is a perfect "home" for positive psychology, as the military consists of relatively young, healthy and pathology-free individuals. In a positive psychology framework, one would consequently focus on those factors that improve and strengthen the mental health of SANDF soldiers.

In the same vein and in support of what was mentioned elsewhere in this chapter, Kasser[47] indicates that rather than assuming that a good life is defined by the absence of psychopathology, many of the leaders in positive psychology have argued that PWB is a construct embedded in positive psychology, to be studied in its own right.[48] It is about having a proactive stance towards achieving optimal emotional, physical and mental well-being through self-acceptance, personal growth, having a purpose in life, mastering the environment, autonomy and having positive relations with others.[49] Furthermore, PWB has been identified as a predictor of risk behavior[50] that is closely linked to an individual's job security and career behavior. When an individual experiences job insecurity he or she is likely to experience anxiety and stress, which will make the person less likely to control and deal with challenging situations through his or her coping mechanisms. Ruini[51] observes that PWB plays a buffering role in coping with stress and a protective role in mental health. It would thus be safe to postulate that "good" PWB will lessen psychological risk, as conceptualized above, in a soldier.

From what has been stated elsewhere, one can deduce what it means for a soldier to run minimal psychological risk for deployment in the military context, but the question arises how it can be understood and applied in the PWB context. An overview of the positive psychology literature and more specifically the literature dealing with PWB highlights the work of Rothmann and Cooper[52], who designed an integrative diagnostic model for PWB in the work context. Closer analysis of this model suggests that it might have good utility in the context of fitness-for-duty evaluations in the military, especially when it is related to the above-mentioned definition of psychological risk. Therefore this model is explored in more detail while also being contextualised to the military environment into what the author refers to as an integrative military model for soldiers' psychological well-being in the external deployment context.

Integrative Military Model for Soldiers' Psychological Well-Being in the External Deployment Context

Rothmann and Cooper's[53] model is designed to address PWB in the work context, which in this instance would be the military environment in general and

A Proposition for the South African National Defence Force 117

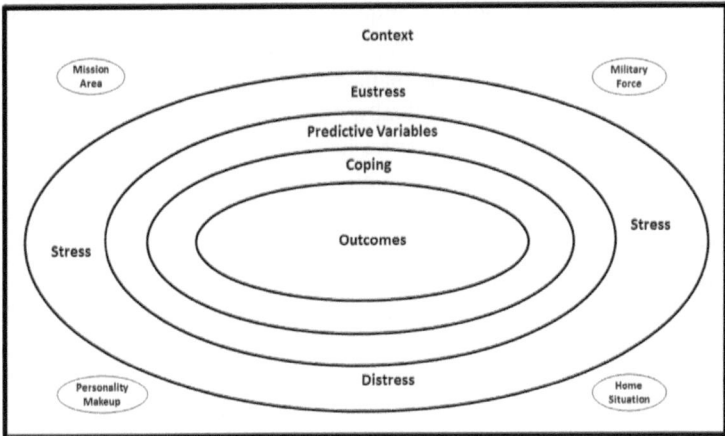

Figure 1: Integrative model for soldier psychological well-being in the external deployment context.

more specifically deployments outside the borders of South Africa, also referred to as external deployments. See Figure 1[54] below for a graphic representation of the suggested integrative military model.

The context that is illustrated by the outer square in Figure 1 will typically refer to the anticipated mission area as a social system with everything related to it, in combination with the physical characteristics of the country, for example whether it is a desert or tropical forest, with its weather patterns and all other factors related to it. The military force (work context) includes own and opposing forces, the deploying soldier's situation back home in terms of his or her family, the support systems, and what is happening in his or her home unit. The context will also include the personality makeup of the soldier, which would be his or her personality, psychological resources and interpersonal relations as manifested in terms of group dynamics.

Moreover, a number of "outside forces"[55] are present, such as social change, race, gender, social class and community, family and other environmental factors; plus "organizational forces" such as job demands, and job resources. As different soldiers tend to react differently to the same stressor, there are also moderating forces such as perception, job experience, self-efficacy, social support, locus of control, sense of coherence, psychological hardiness, coping and optimism that contribute to the manifestation of stress. In some cases, it will be eustress and in others it will be distress. Cilliers and Flotman[56] observe that distress due to non-coping manifests in negativity, emotional alienation, an increase in bureaucratic and autocratic leadership, poor decision-making and ineffective process and people management, which will all have an adverse impact on deployed forces. Individually, or in combination, these forces affect the well-being of soldiers, manifesting in physical effects such as headaches, ulcers, insomnia and decreased immunity; psychological effects such as anger, depression, tension and boredom; behavioral effects such as substance abuse, overeating or undereating, and sleeplessness.[57] From these forces one can formulate predictive variables from which one can predict the outcomes in relation to the particular soldier's ability to cope with these stressors. These predictive variables can be used to make predictions of soldiers' future behavior based on present and past behavior.

From the discussion above the question arises how this positive psychology related model can be applied to the fitness-for-duty evaluation process in the SANDF. Below follows a proposal on a process that can be followed.

Proposed Pre-Deployment Fitness-For-Duty Evaluation Process Based on Positive Psychology

It is evident that in the current process the focus is stronger on a medical model of decision-making with respect to whether a soldier is fit for duty outside South Africa. It is therefore suggested that the proposed process should move away from the stigmatized medical process and focus more strongly on risk assessment by looking at what in the literature is referred to as "the whole person".[58] The medical process is only part of a more comprehensive process, which views the individual soldier as a complex being influenced by a variety of factors in and outside the workplace. The medical model will, for example, not be able to identify certain types of dysfunctional behavior, (e.g., if the person is dishonest or a bully).

The main idea is to obtain as much information as possible to be able to make a decision that a person is an acceptable psychological risk for external deployments. For this reason, various role players are included to become part of the team, whose main aim would be to make an informed decision on whether a particular soldier is suitable for external deployment. Such an informed decision can be made through weighing up information related to his or her medical status and general behavior in relation to the context in which he or she is functioning. In this regard De Soir[59] observes that if a soldier's pre-deployment well-being is relatively low, it is likely to decrease further during the period of deployment.

Furthermore, it is suggested that the new decision-making process should function in the format of a panel, which is convened after the units and individuals earmarked for deployment have been identified and after all relevant medical and other information has been obtained. This can be quite an easy process if the required information is obtained, as in most cases it is a whole unit, such as an infantry battalion, that will be deploying, and it is usually the support staff that come from other units. In the large infantry units, it would be easy to involve immediate supervisors in the panel, but it might not be possible for support personnel from other units. In practice it means that if a specific platoon or company is deployed, that whole company and/or platoon is dealt with simultaneously at the panel. The objective of the panel is to identify those soldiers deemed to be imminently or potentially dangerous, thus posing a risk to themselves and to others. Health information provided by health care practitioners (medical, psychological, social work, and other relevant medical experts), is supported by the inclusion of collateral information from command, the military legal practitioner, the spiritual guide (chaplaincy)[60], and counter-intelligence personnel. The inclusion of command refers to the person's immediate superior. Budd and Harvey[61] support inclusion of the immediate superior and senior enlisted personnel because they describe them as usually the best sources of information on a soldier's behavior and work performance, followed by the person him- or herself, the person's records and lastly psychological testing.

It is furthermore suggested that the person designated to command the external deployment be the chairperson of this panel. However, care should be taken not to divulge confidential medical information on the person, but to restrict inputs to the color code. Any information that the role players think ought to be considered for the decision should be mentioned at the panel. The military law officer can, for example, confirm whether the person has a record of misconduct or if he or she has any outstanding court cases. Based on the author's personal experience, many of the soldiers who were repatriated from external mission areas for conduct-related issues were already a challenge for their commanders in their home country. This kind of information can be considered to obtain a holistic picture of the "whole person", not only the medical information. In this way individuals that already present a challenge would be excluded from external deployment because of the psychological risk that they pose, based on the notion: "… *past behavior predicts future behavior* …"

This is a new process, although it may combine currently existing processes such as confirming whether a person has any outstanding legal issues that function separately from one another. In this way all the information would be made available in one place (under one umbrella) and an informed decision could be taken under the chairmanship of the commander who has to command the force during the external deployment. The concern might be voiced that it might be a long and complicated process, but if all role players arrive with the required information, it is only those who pose a significant risk that will require some time for deliberations, as opposed to the others who are likely to require a few minutes of discussion to make a decision on whether they may deploy externally or not. The benefit of an informed decision that could not have been made in the past will turn the scale in favor of such an integrated process. Information obtained from the panel discussions can also be used to determine trends and individuals who might require interventions. It might even be possible to identify what positive psychology related interventions can be made as part of preparing the force for the particular mission.

This suggested process is theoretical in nature and still needs to be tested in practice in juxtaposition with the foreseen role players. However, it is a good start and when considering that the founding father of social psychology, Kurt Lewin, surmised that "*There is nothing so practical as good theory*"[62] it implies that a good foundation has already been laid for a process that can possibly work. Therefore, it would be apt to make some recommendations to military decision-makers on the proposition to approach pre-deployment fitness-for-duty evaluations from a positive psychology perspective.

Recommendations for Military Decision-Makers

Based on the above discussion, it is thus suggested that military decision-makers in the SANDF in general and more specifically in the SAMHS move away from the stigmatized view of psychopathology to that of psychological risk. From a positive psychology approach, psychological risk can be identified more effectively, and it can be approached as something more manageable, like any other workplace safety and health risk. In this way commanders would also have more control over managing the risks. It will however remain very important to identify those soldiers deemed to be imminently or potentially dangerous in order firstly to minimize the risk posed to themselves or others and secondly to be able to assist the member in his or her well-being.

The importance of the role of leadership in managing the psychological well-being of subordinates is emphasized by Rothmann and Cooper[63] when they suggest that "good" leaders must initiate a process to assess the impact of the various risk factors on overall functioning at present and in the future. It is also important that leadership identify what can be done to improve the overall well-being of soldiers individually or/and as part of the "military family" to enable them to function optimally and to be able to be utilized in the military in fulfilling its constitutional obligations.

It is crucial that military decision-makers initiate and support further research into pre-deployment screening processes and into coping under deployment situations with their unique stressors. This can also include research into identifying soldiers' existing coping mechanisms and then developing programs to improve their coping skills prior to and during deployments. Closely linked to these programs is training that can be contained in employee well-being programs that address issues such as time-, stress- and self-management and coping strategies, addressing personality characteristics such as psychological hardiness, resilience and personal effectiveness. In this way areas for interventions such as psychoeducation can assist in improving a soldier's general PWB. Positive psychology-based interventions can also provide effective tools for working with psychologically distressed soldiers.[64] Military psychologists can thus follow a positive psychology approach in screening soldiers for deployment and can consider looking into post-deployment programs for reintegrating soldiers into their communities and non-deployment routines, as this is essential to lay the foundation for future deployments. The crux of the recommendations made to military decision-makers is to make a deliberate effort to include positive psychology into the regime of the SANDF to improve the PWB of soldiers.

Conclusion

A positive psychology approach will not mean a move away from "selecting out". It is however a "tool" in the hand of the military psychologist where the current selection instrument becomes part of a more inclusive and more informative process of considering collateral information in deciding on the suitability of individuals for external deployments. It is likely to be more advantageous for the SANDF and it makes the deploying force commander part of the decision-making process. In this way the force commander has more information on which informed decisions can be made than in the past and more useful and succinct feedback can be given to sub-level commanders. This will give the clients of the SAMHS more insight into the process that was followed to reach a decision; the client (in this case the force commander) will become part of making the decision. The client can take responsibility for assisting with the development of the affected soldier's PWB and will have a better understanding of what might be required to ensure the PWB of other soldiers as well.

Notes

1 This is an academic document and contains facts and opinions that the author alone considered appropriate and correct for the subject. It does not

necessarily reflect the official policy or the opinion of the South African Government or Department of Defence and Military Veterans or the University of Stellenbosch. Furthermore, a reviewed and completely reworked version of this article is published in the South African Journal of Military Studies.

2 Peter Fabricius, "SA should first sort out its own backyard." *Institute for Security Studies*. 20 July (2017), accessed March 2019, https://issafrica.org/amp/iss-today/sa-should-first-sort-out-its-own-backyard.

3 Petrus C. Bester and John W. O'Neil, "The military leadership training landscape of the South African National Defence Force as related to extreme situations." *Manuscript Submitted for Publication* (2019): 1–25.

4 Henri Boshoff, "African battle space and conflict." *Presentation by the Institute for Security Studies* (Undated). Pretoria: Institute for Security Studies, accessed August 2019, http://aardvarkaoc.co.za/wp-content/Conf_Aug_2009/Henri%20Boshoff%202.pdf; Ivor Ichikowitz, "The African battle space is not the Middle East." *Defence Web*. 7 December (2017), accessed July 2019, https://www.defenceweb.co.za/land/land-land/the-african-battle-space-is-not-the-middle-east-ichikowitz/; Nelson Alusala, "Managing the battle space: Women on the frontline in eastern DRC." *Institute for Security Studies*. 14 January (2016), accessed August 2019, https://issafrica.org/research/central-africa-report/managing-the-battle-space-women-on-the-frontline-in-eastern-drc; Bester and O'Neil, "The military leadership training landscape of the South African National Defence Force as related to extreme situations," 1–25.

5 Jacques J. Gouws, "Military stress and resillience, post-psychological assessment: Factors affecting armed forces' resillience." In Alister Macintyre, Daniel Lagace-Roy and Douglas R. Lindsay (Eds.), *Global views on military stress and resillience*, pp. 113–126 (2017). Kingston: Canadian Defence Academy Press.

6 George L. Lindenfield, George Rozelle, John Hummer, Michael R. Sutherland, and James C. Miller, "Remediation of PTSD in a combat veteran: A case study," *NeuroRegulation*, 6, no. 2 (2019): 102–125, https://doi.org/10.15540/nr.6.2.102

7 Ibid., *"Remediation of PTSD in a combat veteran: A case study,"* 102–103.

8 United Nations. *Medical support manual for United Nations field operations* (1995). Department of Peacekeeping Operations.

9 Arthur Neale, "Reflecting on 15 years of psychological (pre- and post-deployment) screening in the SANDF." *Presentation at the 56th Annual International Military Testing Association Conference Held in Hamburg Germany from October 27 to October 31 2014*. Hamburg: IMTA, October, 2014.

10 Ibid., *"Reflecting on 15 years of psychological (pre- and post-deployment) screening in the SANDF."*

11 Roberto J. Rona, Kenneth C. Hyams, and Simon Wessely, "Screening for psychological illness in military personnel." *The Journal of the American*

Medical Association, 293(10) (2005): 1257–1260, https://doi.org/10.100/Jama.293.10.1257
12 Frank C. Budd and Sally Harvey, "Military fitness-for-duty evaluations." In Carrie H. Kennedy and Erica A. Zillmer (Eds.), *Military psychology: Clinical and operational applications*, p. 46 (2006). London: The Guilford Press.
13 Corey L.M. Keyes, Barbara L. Fredriksen and Nansoon Park, "Positive psychology and the quality of life." In Kenneth C. Land, Alex C. Michalos and Joseph M. Sirgy (Eds.), *Handbook of social indicators and quality of life research*, p. 99 (2012). Amsterdam: Springer Netherlands.
14 Military Psychological Institute [MPI]. *Manual for the psychological risk indicator 2.5: A psychological screening tool for concurrent health assessment* (2019). Pretoria: Military Psychological Institute.
15 Keys et al., *"Positive psychology and the quality of life."* 101–102.
16 South African Military Health Service [SAMHS], "South African Military Health Service patient administration policy and procedures." In *SAMHS order: Pat admin 05/2002* (2006). Pretoria: SAMHS.
17 Budd and Harvey, "Military fitness-for-duty evaluations," 35
18 SAMHS, *"South African Military Health Service patient administration policy and procedures,"* 8–5.
19 MPI. *Manual for the psychological risk indicator 2.5: A psychological screening tool for concurrent health assessment*, 1–2.
20 Ibid., *Manual for the psychological risk indicator 2.5: A psychological screening tool for concurrent health assessment*, 16.
21 Ibid., *Manual for the psychological risk indicator 2.5: A psychological screening tool for concurrent health assessment*, 2.
22 SAMHS, *"South African Military Health Service patient administration policy and procedures,"* 8–2.
23 David Sue, Derald W. Sue, Diane M. Sue and Stanley Sue. *Essentials of understanding abnormal behavior*, 3rd ed., 2 (2017). Boston, MA: Cengage Learning.
24 Scott O. Lilienfeld, Sarah F. Smith, and Ashley L Watts, "Issues in diagnosis: Conceptual issues and controversies." In W. Edward Craighead, David J. Miklowitz, and Linda W. Craighead (Eds.), *Psychopathology: History, diagnosis and empirical foundations*, pp. 1–35 (2017). San Francisco, CA: Wiley.
25 European Agency for Safety and Health at Work. *Psychological risk and stress at work* (2019), https://osha.europa.eu/en/themes/psychosocial-risks-and-stress
26 Marise Born and Henk van der Flier, "Risk management through the psychological looking glass." *Youtube*. Video File. May 6, (2019), https://youtu.be/BWVY4H2pNS8
27 Walther Wiese, *Presentation on espionage case studies* (1997). Pretoria: Defence Intelligence Media Centre.

28 Petrus C. Bester, *Background to comprehensive profiling services*, p. 1 (2006). Pretoria.
29 Bernadette McSherry, "Risk assessment by mental health professionals and the prevention of future violent behaviour." *Trends and issues in criminal justice* (July 2004). Canberra: Autralian Institute of Criminology, accessed August 10, 2019, http://www.aic.gov.au.
30 MPI, *Manual for the psychological risk indicator 2.5: A psychological screening tool for concurrent health assessment*, 2.
31 Ibid., *Manual for the psychological risk indicator 2.5: A psychological screening tool for concurrent health assessment*, 2.
32 Jelana Milnič, "Salutogenetic model of health, a sense of family coherence and proactive coping among adolescents." In Anita Pešič (Ed.), *Stress in military profession*, pp. 113–124, 114 (2018). Belgrade: Odbrana.
33 Ibid., *"Salutogenetic model of health, a sense of family coherence and proactive coping among adolescents,"* 114.
34 Carl Jacob and Daniel Lagacé-Roy, "Military stress and resilience: Introduction to the 2017 International Military Testing Association inaugural volume." In Allister MacIntyre, Daniel Lagace-Roy and Douglas R. Lindsay (Eds.), *Global views on military stress and resilience*, pp. 1–16, 3–4 (2017). Kingston: Canadian Defence Academy Press.
35 Ibid., *"Military stress and resilience: Introduction to the 2017 International Military Testing Association inaugural volume,"* 4–5.
36 Ian Rothmann and Cary L. Cooper. *Organizational and work psychology*, 239 (2008). London: Hodder Education.
37 European Agency for Safety and Health at Work. *Psychological risk and stress at work.*
38 Ibid., *Psychological risk and stress at work.*
39 S. Rothman, K. Mostert and M. Strydom, "A psychometric evaluation of the job demands-resources scale in South Africa." *South African Journal of Industrial Psychology*, 32, no. 4 (2006): 76–86.
40 Stephen Joseph, "Preface." In Stephen Joseph (Ed.), *Positive psychology in practice: Promoting human flourishing in work, health, education and everyday life*, pp. xi–xiii (2015). New Jersey: Wiley.
Frans Cilliers and Aden-Paul Flotman, "The psychological well-being manifesting among master's students in industrial and organisational psychology." *SA Journal of Industrial Psychology*, 42 (2016), http://dx.doi.org/10.4102/sajip.v42i1.1323
41 Michael D. Matthews, "Cognitive and non-cognitive factors in soldier performance" In Janice H. Laurence and Michael D. Matthews (Eds.), *The Oxford handbook of military psychology*, pp. 197–217 (2012). London: Oxford University Press.

42 Milnič, "Salutogenic model of health, a sense of family coherence and proactive coping among adolescents," 113–124.
43 Conceptualized by Aaron Antonovsky the salutogenetic paradigm refers to looking at health and illness not as a dichotomy but as a continuum. It focuses rather on the study of the origins of health, opposed to the origins of disease.
44 Nadia Ferreira, "Hardiness in relation to organisational commitment in the human resource management field." *SA Journal of Human Resource Management*, 10, no. 2 (2012), http://dx.doi.org/10.4102/sajhrm.v10i2.418
45 Michael D. Matthews, *Towards a positive military psychology*, Vol 3, in *Military psychology*, Michael D. Matthews and Janice H. Laurence (Eds.), pp. 283–292 (2012). London: SAGE.
46 Ibid., *Towards a positive military psychology*.
47 Tim Kasser, "The science of values in the culture of consumption." In Stephen Joseph (Ed.), *Positive psychology in practice: Promoting human flourishing in work, health, education, and everyday life*, pp. 83–102 (2015). New Jersey: Wiley.
48 Cilliers and Flotman, "The psychological well-being manifesting among master's students in industrial and organisational psychology."
49 Ibid., "The psychological well-being manifesting among master's students in industrial and organisational psychology."
Ian Rothmann and Cary L. Cooper, *Organizational and work psychology*, p. 230 (2008). London: Hodder Education.
50 Ferreira, "Hardiness in relation to organisational commitment in the human resource management field."
51 Chiara Ruini, *Positive psychology in the clinical domains: Research and practice*, p. 45, 84 (2017). Bologna: Springer.
52 Rothmann and Cooper, *Organizational and work psychology*, 229–248.
53 Rothmann and Cooper, *Organizational and Work Psychology*, 229–248.
54 This figure was created by the author of this chapter. It is based on the concepts from Rothmann and Cooper (2008) but contextualized for the military.
55 Cilliers and Flotman, "The psychological well-being manifesting among master's students in industrial and organisational psychology."
56 Ibid., "The psychological well-being manifesting among master's students in industrial and organisational psychology."
57 Rothmann and Cooper, *Organizational and work psychology*, 236–237.
58 Bloom Leadership, "6 Areas of Whole Person Development: Professional." (2017), accessed August 2019, http://www.bloomleaders.com/blog/2017/1/25/6-areas-of-whole-person-development-professional; William Henderson, "Security Clearance: The Whole-Person Concept." (2010), Accessed August 2019, https://news.clearancejobs.com/2010/12/27/security-clearance-the-whole-person-concept/; Ole Boe, Kristin K. Wooley and John Durkin,

"Choosing the elite recruitment, assessment and selection in law enforcement tactical teams and military special forces." In Patrick J. Sweeney, Michael D. Matthews and Paul B. Lester (Eds.), *Leadership in dangerous situations: A handbook for the armed forces, emergency services, and first responders*, pp. 333–349 (2011). Annapolis: Naval Institute Press.

59 Erik de Soir, "Psychological adjustment after military operations: 'The Utility of Postdeployment Decompression for Supporting Health Readjustment.'" In Stephen V. Bowles and Paul T. Bartone (Eds.), *Handbook of military psychology: Clinical and organizational practice*, pp. 89–103 (2017). Cham: Switzerland.
60 This will be omitted if the person has no spiritual connection.
61 Budd and Harvey, *"Military fitness-for-duty evaluations"*.
62 Paul Marsden, *"Nothing so Practical as Good Theory: Syzgy's Mark Ellis on Social Commerce,"* Accessed 17 August, 2019, https://digitalwell-being.org/nothing-so-practical-as-good-theory-syzygys-mark-ellis-on-social-commerce/
63 Rothmann and Cooper, *Organizational and work psychology*, 238.
64 Matthews, *Towards a positive military psychology.*

Nity Sharma and Vrishti Kapoor

Flourish: Promoting Positivism and Psychological Well-Being in Military

> *The gold standard for measuring well-being is flourishing, and that the goal of positive psychology is to increase flourishing*
>
> *Martin E P Seligman[1]*

Introduction

Attainment of optimal psychological well-being has remained from eternity an important domain of research, as well-being is fundamental to individuals' overall health, which is one of the universal human needs. Psychological well-being, conceptualized as an outcome of a pleasant life, has been the object of substantial empirical exploration. Several theorists have used the concept of 'psychological well-being'[2,3] interchangeably with the concept of flourish. Flourish is defined as having high levels of both hedonic and eudaimonic well-being.[4,5,6] According to Keyes,[7] flourish is a combination of social, emotional, and psychological well-being. Although various operationalizations of flourish exist,[8] the most recent one is Seligman's[9] 'PERMA-Profiler' which divides human flourishing into five pillars: Positive Emotion, Engagement, Relationships, Meanings and Accomplishment. According to Seligman, these elements which individuals pursue in their efforts to flourish, represent the fundamentals of human well-being. In his presidential address to the American Psychological Association in 1998, Seligman,[10] who heralded the positive psychology movement, emphasized the importance of well-being over happiness. If positive psychology is to be more than a 'happiology', its focus needs to shift to well-being. The goal of positive psychology is to increase flourishing, which is the gold standard for measuring well-being.[11]

Well-being is very subjective and includes the key components of self-acceptance, personal growth, purpose in life, environmental mastery, autonomy, and positive relations with others.[12] To some extent, psychological well-being (PWB)is relatively stable, and it is influenced by past experiences as well as by personality traits. Stressful experiences influence the levels of PWB whereas positive daily experiences contribute to the maintenance a good level of PWB.

The PWB of military personnel is of utmost importance not only for the armed forces but for the country as well. Military operations expect soldiers to perform in a zero-error context, which puts undue pressure on them as it requires high levels of physical and mental stamina. This explains, at least in part, why the military profession stands among one of the most stressful occupations.[13] The available literature abundantly highlights the psychological issues experienced by military personnel, such as occupational stress,[14] post-traumatic stress disorder (PTSD), anxiety, depression, and alcohol abuse.[15,16] Although empirical evidence suggests that it is important for military personnel to have strong resilience,[17,18] the evidence for a positively oriented approach to promote PWB along with physical well-being other than during the initial training in the military is scarce. Hence, keeping a regular check on military personnel's PWB has become a necessity. Furthermore, from time-to-time, restorative activities would not only improve performance, but also reinforce mental health and overall well-being. Thus, this chapter draws upon one of the emerging dimensions of positive psychology for promoting PWB and positivism and discusses implementation of its five pillars in the military context.

Psychological Well-Being

The scientific approach to the study of PWB has historical roots. The Sankhya philosophy describes PWB as an equilibrium or homeostasis that individuals seek to achieve in order to experience well-being.[19] In his classic work, Bradburn defined well-being in terms of positive and negative effects, inspired by Aristotle's philosophical orientations: hedonic (presence of positive affect and absence of negative affect) and eudaimonic (full psychological actualization).[20] Over time, the concept of well-being has evolved in terms of its dimensions. Eminent researchers attempted to define the structure of this concept by emphasizing various elements. Influenced by Rogers' conceptualization of a fully functioning person, Carol Ryff elaborated upon the following aspects of well-being: autonomy, self-acceptance, positive relationships, environmental mastery, personal growth, and purpose in life.[21,22,23] Others attempted to provide a concise definition of well-being by equating it with a global assessment of an individual's quality of life.[24] The eminent psychologist Ed Diener viewed well-being as a subjective construct as he proposed that people evaluate their own degree of wellness based on three distinct components: higher positive affect, lower negative affect, and a cognitive evaluation of satisfaction with life.[25] The self-determination theory formulated by Edward Deci and Richard Ryan argues that competency, autonomy, and relatedness are the three essential components of PWB.[26] However, the most recently drawn conceptualization of PWB by Martin Seligman, flourish, provides a holistic perspective, which we expound in the following section.[27]

Flourish

The most widely accepted definition of health provided by the World Health Organization in 1946, describes it as "a state of physical, mental and social well-being, and not merely an absence of disease."[28] Thus, PWB is equally as important as physical fitness. Although Seligman initially defined PWB in terms of 'happiness,' he soon corrected himself by adding that it constitutes much more than 'just being happy.'[29] He thus talked about *Flourishing*, a refined indicator of well-being, which rests on five pillars, denoted by the acronym PERMA. PERMA includes five core elements of well-being, namely positive emotions, engagement, relationships, meaning, and accomplishments.[30] Positive emotions contribute to the ability to view situations from an optimistic perspective, which is different from simply being happy. Engagement describes one's complete absorption in activities of high subjective interest, which contribute to personal and professional growth. Developing deep and meaningful social relationships, which can provide support in stressful situations, is another crucial element that contributes to overall well-being. Identifying a meaning and purpose in life strengthens and justifies one's choices and decisions. Finally, achievement of one's goals and ambitions can lead to a sense of satisfaction and contentment.

Flourish and Well-Being

In the last three decades, the conceptual literature on well-being has revolved around two distinct philosophical traditions, hedonia and eudaimonia.[31,32] The hedonic view of well-being refers to a state of pleasure and happiness.[33,34] In contrast, the eudaimonic view conceptualizes well-being in four major ways: gaining personal strengths and contributing to the greater good,[35] acting in accordance with one's inner nature and deeply held values,[36] realizing one's true potential;[37] and living a purposeful or meaningful life.[38] In other words, hedonia can occur through basic sensory pleasures triggered by food, physical arousal, or social or physical activities. However, eudaimonia stems from a cognitive perception of morally living well, in which individuals feel that their life is valuable, meaningful, and engaging. Hedonia has been considered synonymous with subjective well-being, wherein humans essentially desire to maximize their experience of pleasure and to minimize pain, while eudaimonia constitutes a more objective kind of well-being, in which the quality of one's life is judged according to the extent to which it is a life of excellence and virtue.[39] Recent years have witnessed some debate regarding the distinctions between hedonia and eudaimonia.[40] Debate aside, research evidence suggests that a life rich in both types of pursuits is associated with the highest degree of well-being. Thus, researchers acknowledge the importance of both the hedonic and eudaimonic approaches, the combination of which has given birth to the term flourish.[41,42] Thus, the concepts of well-being and flourish are intricately integrated and lead to overall well-being.

Manifestations of Flourish

Flourish manifests into a wide variety of healthy physical and psychological outcomes. It broadens behavioral repertoires and increases intuition, creativity and pleasant feelings. Benefits can manifest physiologically, such as healthy indices of cardiovascular functioning, like lowered blood pressure. Also, positive affect and flourish are related to longevity.[43] In addition to predicting mental and physiological benefits, flourish is associated with increased self-efficacy and pro-social behavior, which in turn leads to higher motivation to work actively and to pursue new goals. This helps people attain life and societal goals, and perform optimally in the workplace, which results in higher satisfaction and reinforces more positivism, thereby creating an upward spiral.[44] Huppert and So[45] conceptualized flourish as the experience of "life going well." It is a combination of feeling good and functioning effectively and is synonymous with a high level of mental well-being. Flourishers tend to experience better mental health and are more resilient to life stressors than non-flourishers.[46,47,48,49] Keyes[50] categorized individuals who flourish as being free of mental disorders, moderately mentally healthy, and not languishing.Identifying the characteristics of highly flourishing individuals can inform researchers, psychologists and policymakers who seek to improve people's ability to flourish.[51]

Challenges to Well-Being in a Military Context

Military life requires discipline. The transition or adaptation from the civilian life to a military service presents several psychological challenges.[52] Basic military training, purports to develop self-discipline, physical fitness, confidence, loyalty to and pride in serving the country, and other essential military values.[53] However, it has also been found to induce a sense of loss of personal control and disappointment,[54] elevated anxiety levels during the initial period and sleep disturbance,[55] fluctuation in mood and stress levels, and self-reported psychological distress.[56]

Soldiers also encounter deployment-related stressors or severe psychological outcomes as military life exposes them to challenging and threatening situations.[57] Separation from family, working continuously for long hours, and the uncertain nature of deployment, taken together, tend to test interpersonal relationships, one's intention to stay in the military, and induce higher stress levels.[58] Soldiers are often exposed to traumatic events like death and/or injury, handling human remains, assisting the wounded, and being fired on, which often lead to serious post deployment mental health issues such as anxiety, depression, substance abuse, decreased satisfaction with life, adjustment disorders, and aggression.[59]

Work stress has the potential to disrupt work performance. Pflanz and Ogle, concluded that work stress represents a significant occupational hazard in the military work environment, leading to impaired work performance, increased missed workdays, negative perception of superiors, and deteriorating physical health.[60] Colonel K C Dixit[61] identified major triggers of stress among military officers which may be divided into two broad categories: professional stressors and domestic stressors. Professional stressors comprise interpersonal conflicts at work, shortage of officers, responsibilities overload, poorly articulated tenure policies, and unfulfilled career and life ambitions. Domestic stressors include the inability to satisfy their children's desires, denial of leave at crucial times, shortage of married accommodations and lack of basic amenities, and imbalance in meeting professional versus familial demands and responsibilities. Taken together, these stressors lead to significantly reduced satisfaction levels, missed promotions, personnel opting for premature retirement, a sense of ambiguity in operations, loneliness, casualties, battle fatigue, and other significant operation-related psychological consequences.

Air combat operation stressors that can lead to various negative emotional reactions:-include the uncertain nature of operations, anticipation, exposure to a hostile environment, reduced availability of personnel or resources, confirmed casualties of war, resettlement, rehabilitation of injured soldiers, and financial instability.[62] The resultant emotional reactions comprise shock, irritability, fear, guilt, anxiety, sadness, tension for family members, stress, worry about deployed members, etc. These, in turn, may also lead to reduced performance quality and unhealthy behaviors like substance abuse. Similarly, soldiers previously exposed to counter insurgency operations reported experiencing sleep disturbances, anxiety, depression, and conflict with superiors, all linked to various stressors like direct combat exposure, operational stress, or concerns about the home front.[63]

Non-deployment factors, such as work relationships, role conflict, imbalance between efforts and rewards, excessive job demands, work-life imbalance, and financial issues also affect military personnel's PWB.[64] There is a positive relationship between perceived organizational support and service members' mental health, implying that negatively perceived organizational support and heightened stress are related to negative mental health outcomes such as deteriorated well-being and increased risk of PTSD, along with increased stigma for seeking treatment for mental health issues.[65]

Thus, a plethora of scientific evidence corroborates that military life throws a huge array of challenges for service members to endure. These stressors may negatively influence job satisfaction, intensity of work engagement or involvement, the quality of work interpersonal relationships as well as family relationships, perceived meaning of life, perception of accomplishments, and experienced emotions, ultimately straining soldiers' PWB. The point of concern is that preparation for and rehabilitation after deployments include comprehensive physical training to ensure sound physical health, but mental health is hardly discussed, thereby perpetuating the existing stigma.[66] Hence, it is essential to intervene with the purpose of enhancing soldiers' PWB, to augment work performance, which is necessary to be able to protect competently a nation.

Flourish and Psychological Well-Being in Military Context

The detrimental impact of military stressors is mediated by certain protective factors like personality characteristics, sources of support, and training.[67] Thus, the strengthening of such factors can help in improving as well as sustaining an optimal level of PWB, which is an essential requirement in the military. On this account the flourish framework is relevant in several ways.

The first building block of flourish, *positive emotions*, is an intangible but salient component in the life of military personnel. It relates soldiers' resilience to fight during traumatic situations, as suggested by a correlation between positive affect, mood states, and cognitive appraisals with improved psychological adjustment during traumatic conditions.[68] Strong positive correlations have emerged between positive emotions, higher resilience and improved psychosomatic health, and similarly between negative emotions, lower resilience and deteriorated psychosomatic health.[69] Thus, it would be safe to claim that 'feeling good' may be important to 'be good' at work.

The second element of flourishing, *engagement*, may be used as an index of military personnel's job satisfaction. Allowing soldiers to be creative and strategic in how they carry out their tasks and assigning them responsibilities, may facilitate their complete their complete engagement in their military career, leading to enhanced interest and self-worth. For instance, emotional exhaustion, decreased intrinsic motivation, and dissatisfaction with salary and promotion opportunities are known predictors of job turnover in the military.[70] Such factors are likely to affect negatively military personnel's work engagement level and thus, need regular monitoring. Highly motivated and engaged teams serve as buffers against organizational constraints during deployment, which results in fewer fatigue symptoms post deployment,[71] and better team performance. Life stress, a stressful work environment, task-related stress, organizational stress, as well as stress induced by inactivity during idle time, all contribute to increase stress levels, impact personnel' job satisfaction and further affect soldiers' operational effectiveness.[72] Hence, the introduction of educational, research and recreational activities, as well as life and organizational skills training, may be a few tools that could help in reducing soldiers' stress.[73]

The third component of flourishing, *meaningful relationships*, may be considered the backbone of individuals as well as a strong contributor to resilience. Unit support from superiors and fellow unit members, and social support from friends and family, especially during and post deployment periods, constitute a strong buffer against soldiers' PTSD.[74] Furthermore, trait resilience along with unit cohesion and social support possibly reduce negative cognitions, which mitigate the severity of PTSD.[75] The buddy system in the armed forces provides a personal source of support that helps in dealing with deployment stressors[76] and enhances the. *esprit de corps* in a unit. Moreover, meaningful relationships may also facilitate the expression of positive as well as negative emotions, thereby reducing the suppression of emotions such as feelings of loneliness and longing for family that deployed soldiers find very stressful.[77]

Conversely, threatened, or strained interpersonal relationships, with team members, family, and/or friends, may lead to enhanced stress, especially during crucial times like that of deployment. Women veterans identified interpersonal issues as the most commonly experienced stressor during deployment.[78] Thus, the presence of meaningful relationships is indispensable in a demanding profession like the military and therefore, military group cohesion is of utmost importance, especially during personnel training.

The fourth element of flourishing, *meaningfulness in life* comprises the meanings that soldiers give to their profession, which in turn motivates them to put themselves continually in danger for others' safety. Selection personnel can help recruits in finding meaning or purpose in their choice of military occupation. For instance, the motto of the Indian Army, 'Service before Self' reminds soldiers of the ultimate *meaning and purpose* for their voluntary enrolment, leading to constant motivation. A literature review concluded that deployed personnel constantly draw motivation to serve in challenging missions from the meaningfulness that they derive from a sense of being able to help and support others.[79] This perceived meaningfulness further fosters subjective personal strength, work, self-fulfillment, and self-reliance, as revealed by a qualitative study consisting of an in-depth interview of 19 veterans.[80] In contrast, an increased level of post deployment stress is associated with a subjective lack of meaning of work activities.[81] Having a stronger meaning in life helps reduce the severity of emotional distress and suicide risk, and promotes performance and success across multiple domains in solders' life.[82] Thus, the available literature supports and strengthens the necessity of a stronger meaningfulness or purpose in life, as a contributing factor that enhances soldiers' PWB, as they face traumatic environments.

Finally, the fifth building block of a flourishing life, *achievement*, is also essential. Achievements tend to buffer the challenging nature of the military profession. Highlighting major as well as minor recruits' *achievements* during their training may help inculcate a sense of satisfaction and enhance recruits' self- confidence.[83] Appreciation of efforts is essential to recognize soldiers' achievements, because reportedly, achievements tend to minimize the ill effect of long working hours, illegitimate tasks, and enhance job satisfaction and well-being.[84]

Thus, the available evidence strongly substantiates that all the five building blocks or components of flourish, in the military context, must be cultivated and reinforced to the same extent as physical trainings to help military personnel achieve an optimum level of performance.

Conclusion

Promoting positivism in the military, with respect to the five essential dimensions of flourishing would help military personnel not only to enhance their work performance but also to live a positive life, which would optimize their mental health and PWB. Based on the theoretical PERMA model, namely Positive emotions, Engagement, Relationships, Meaning, and Accomplishments, this chapter highlighted the association between flourishing, one of the significant concepts of positive psychology and PWB for promoting positivism and well-being in the military.

Recommendations

According to Seligman,[85] to flourish, individuals must possess characteristics considered essential (positive emotions, engagement, interest, meaning and purpose). Such characteristics can be developed and constantly improved and nurtured through training not only at the beginning of one's military career, but throughout one's entire military career. This would not only lead to achievement of PWB but also would have far-reaching consequences in terms of enhanced performance, increased resilience, heightened motivation, and overall optimum well-being. The long-term goal is to develop strategies that would support flourishing among military personnel, thereby improving their PWB and overall quality of life.

Notes

1 Martin Seligman, "Authentic happiness." *Flourish: A new theory of positive psychology (Archived Newsletter)*, 2011. Retrieved from https://www.authentichappiness.sas.upenn.edu/newsletters/flourishnewsletters/newtheory
2 Richard Ryan and Edward Deci, "On happiness and human potentials: A review of research on hedonic and eudaimonic well-being." *Annual Review of Psychology*, 52, no. 1 (2001): 141–166, https://doi.org/10.1146/annurev.psych.52.1.141
3 Carol Ryff, "Happiness is everything, or is it? Explorations on the meaning of psychological well-being." *Journal of Personality and Social Psychology*, 57, no. 6 (1989): 1069–1081, https://doi.org/10.1037/0022-3514.57.6.1069
4 Felicia Huppert and Timothy So, "Flourishing across Europe: Application of a new conceptual framework for defining well-being." *Social Indicators Research*, 110, no. 3 (2013): 837–861, https://doi.org/10.1007/s11205-011-9966-7
5 Felicia Huppert, "Psychological well-being: Evidence regarding its causes and consequences." *Applied Psychology: Health and Well-Being*, 1, no. 2 (2009): 137–164, https://doi.org/10.1111/j.1758-0854.2009.01008.x
6 Corey Keyes, "The mental health continuum: From languishing to flourishing in life." *Journal of Health and Social Research*, 43 (2002): 207–222, https://doi.org/10.2307/3090197
7 Ibid.
8 Lucy Hone, Aaron Jarden, Grant Schofield, and Scott Duncan, "Measuring flourishing: The impact of operational definitions on the prevalence of high levels of well-being." *International Journal of Well-being*, 4, no. 1 (2014): 62–90, https://doi.org/10.5502/ijw.v4i1.4
9 Martin Seligman, *Flourish: A new understanding of happiness and well-being—and how to achieve them* (2011). London: Nicholas Brealey.
10 M. E. P. Seligman (1999).
11 Martin Seligman, "Authentic happiness." *Flourish: A new theory of positive psychology*.
12 Ibid.

13 CareerCast.com, *"2019 most stressful jobs."* 2019. Retrieved from https://www.careercast.com/jobs-rated/most-stressful-jobs-2019
14 Sakshi Sharma, "Stress Management in Indian Army." *Daily Excelsior Magazine*, March 09, 2014. Retrieved from https://www.dailyexcelsior.com/stress-management-in-indian-army/
15 Peter Harms, Dina Krasikova, Adam Vanhove, Mitchel Herian, and Paul Lester, "Stress and emotional well-being in military organizations." *Research in Occupational Stress and Well Being*, 11 (2013): 103–132, https://doi.org/10.1108/S1479-3555(2013)0000011008
16 K.C. Dixit, *Addressing stress-related issues in army* (2011). New Delhi: Institute for Defence Studies and Analysis.
17 John Cacioppo, Harry Reis, and Alex Zautra, "Social resilience: The value of social fitness with an application to the military." *The American Psychologist*, 66, no. 1 (2011): 43–51, https://doi.org/10.1037/a0021419
18 G.W. Casey, "Comprehensive soldier fitness: A vision for psychological resilience in the US Army." *The American Psychologist*, 66, no. 1 (2011): 1–3, https://doi.org/10.1037/a0021930
19 Christie Kelley, Thomas Britt, Amy Adler, and Paul Bliese, "Perceived organizational support, posttraumatic stress disorder symptoms, and stigma in soldiers returning from combat" *Psychological Services*, 11, no. 2 (2013): 1–6, https://doi.org/10.1037/a0034892
20 Sinha Durganand, "Concept of psycho-social well-being: Western and Indian perspectives" In Ajit Dala and Girishwar Misra (Eds.), *New directions in health psychology* (2011). New Delhi: Sage Publications.
21 Carmelo Vazquez, Gonzalo Hervas, Juan Jose Rahona, and Diego Gomez, "Psychological well-being and health. Contributions of positive psychology." *Annuary of Clinical and Health Psychology*, 5 (2009): 15–27. url: https://www.researchgate.net/publication/228460254_Psychological_well-being_and_health_Contributions_of_positive_psychology
22 Carol Ryff, *"Happiness is everything, or is it? Explorations on the meaning of psychological well-being."*
23 Carol Ryff, "Psychological well-being in adult life." *Current Directions in Psychological Science*, 4, no. 4 (1995): 99–104, https://doi.org/10.1111/1467-8721.ep10772395
24 Rachel Dodge, Annette Daly, Jan Huyton, and Lalage Sanders, "The challenge of defining well-being." *International Journal of Well-being*, 3 (2012): 222–235, https://doi.org/10.5502/ijw.v2i3.4
25 Edward Deci and Richard Ryan, "Hedonia, eudamonia, and well-being: An introduction." *Journal of Happiness Studies*, 9 (2008): 1–11, https://doi.org/10.1007/s10901-006-9018-1
26 Ibid.

27 Dodge, Daly, Huyton, and Sanders, *"The challenge of defining well-being"*.
28 Sowgandhi Chaturvedula and Catherine Joseph, "Dimensions of psychological well-being and personality in military aircrew: A preliminary study." *Indian Journal of Aerospace Medicine*, 51, no. 2 (2007): 17–27. Retrieved from http://medind.nic.in/iab/t07/i2/iabt07i2p17.pdf
29 Martin Seligman, *Flourish: A visionary new understanding of happiness* (2011). New York, NY: Free Press.
30 Martin Seligman, "PERMA and the building blocks of well-being." *The Journal of Positive Psychology*, 13, no. 4 (2018): 1–3, https://doi.org/10.1080/17439760.2018.1437466
31 Martin Seligman, *Flourish: A new understanding of happiness and well-being – And how to achieve them*.
32 Kent Berridge and Morten Kringelbach, "Building a neuroscience of pleasure and well-being." *Psychology of Well-Being: Theory, Research and Practice*, 1, no. 1 (2011): 1–3, https://doi.org/10.1186/2211-1522-1-3
33 Daniel Kahneman, Ed Diener, and Norbert Schwarz. *Well-being: The foundations of hedonic psychology* (1999). New York: Russell Sage Foundation.
34 Richard Ryan and Edward Deci, *"On happiness and human potentials: A review of research on hedonic and eudaimonic well-being."*
35 Aristotle. *Nicomachean ethics*. R. Crisp (Trans.) (2000). Cambridge: Cambridge University Press.
36 Alan Waterman, "Two conceptions of happiness: Contrasts of personal expressiveness (Eudaimonia) and hedonic enjoyment." *Journal of Personality and Social Psychology*, 64, no. 4 (1993): 678–691, https://doi.org/10.1037/0022-3514.64.4.678
37 Carol Ryff and Corey Keyes, "The structure of psychological well-being revisited." *Journal of Personality and Social Psychology*, 69 (1995): 719–727, https://doi.org/10.1037/0022-3514.69.4.719
38 Carol Ryff, *"Happiness is everything, or is it? Explorations on the meaning of psychological well-being"*.
39 McDowell (1980).
40 Luke Wayne Henderson, Tess Knight, and Ben Andrew Richardson, "An exploration of the well-being benefits of hedonic and eudaimonic behaviour." *The Journal of Positive Psychology*, 8, no. 4 (2013): 322–336, https://doi.org/10.1080/17439760.2013.803596.
41 Huppert and So (2009).
42 Luke Wayne Henderson, Tess Knight, and Ben Andrew Richardson, *"An exploration of the well-being benefits of hedonic and eudaimonic behaviour"*.
43 Barbara Fredrickson and Marcial Losada, "Positive affect and complex dynamics of human flourishing." *American Psychologist*, 60, no. 7 (2005): 678–686, https://doi.org/10.1037/0003-066x.60.7.678. PMC 3126111. PMID 16221001

44 Sonja Lyubomirsky, Laura King, and Ed Diener, (in press), "The benefits of frequent positive affect: Does happiness lead to success?" *Psychological Bulletin*, 131, no. 6 (2005): 803–855, https://doi.org/10.1037/0033-2909.131.6.803
45 Huppert and So (2009).
46 Bergsma et al. (2011)
47 Felicia Huppert, *"Psychological well-being: Evidence regarding its causes and consequences."*
48 Rosemarie Kobau, Martin Seligman, Christopher Peterson, Ed Diener, Matthew Zack, Daniel Chapman, and William Thompson, "Mental health promotion in public health: Perspectives and strategies from positive psychology." *American Journal of Public Health*, 101, no. 8 (2011): 1–9, https://doi.org/10.2105/AJPH.2010.300083
49 Ruut Veenhoven, "Healthy happiness: Effects of happiness on physical health and the consequences for preventive health care." *Journal of Happiness Studies*, 9, no. 3 (2008): 449–469, https://doi.org/10.1007/s10902-006-9042-1
50 Corey Keyes, *"The mental health continuum: From languishing to flourishing in life."*
51 Huppert and So (2009).
52 Jeff Cigrang, Sandy Todd, and Eric Carbone, "Stress management training for military trainees returned to duty after a mental health evaluation: Effect on graduation rates." *Journal of Occupational Health Psychology*, 5(1) (2000): 48–55, https://doi.org/10.1037//1076-8998.5.1.48
53 Ibid.
54 Ibid.
55 Martin et al., *"Psychological adjustment during army basic training, 2006".*
56 Can Nakkas, Hubert Annen, and Serge Brand, "Psychological distress and coping in military cadre candidates." *Neuropsychiatric Disease and Treatment*, 12 (2016): 2237–2243, https://doi.org/10.2147/NDT.S113220
57 David Carless, Suzanne Peacock, Jim McKenna, and Carlton Cooke, "Psychosocial outcomes of an inclusive adapted sport and adventurous training course for military personnel." *Disability and Rehabilitation*, 35, no. 24 (2013): 2081–2088, https://doi.org/10.3109/09638288.2013.802376
58 James Hosek, Jennifer Kavanagh, and Laura Miller. *How deployments affect service members* (2006). Santa Monica, CA: Rand Corporation.
59 David Carless, Suzanne Peacock, Jim McKenna, and Carlton Cooke, *"Psychosocial outcomes of an inclusive adapted sport and adventurous training course for military personnel."*
60 Steven Pflanz and Alan Ogle, "Job Stress, depression, work performance, and perceptions of supervisors in military personnel." *Military Medicine*, 171, no. 9 (2006): 861–865, https://doi.org/10.7205/MILMED.171.9.861

61 Dixit (2011). *Addressing stress related issues in the army.*
62 Catherine Joseph, "An overview of psychological factors and interventions in air combat operations." *Indian Journal of Aerospace Medicine*, 51, no. 2 (2007): 1–16. Retrieved from http://medind.nic.in/iab/t07/i2/iabt07i2p1.pdf
63 Larry Applewhite, Nathan Keller, and Adam Borah, "Mental health care use by soldiers conducting counterinsurgency operations." *Military Medicine*, 177 (2012): 501–506, https://doi.org/10.7205/milmed-d-11-00142
64 Sam Brooks and Neil Greenberg, "Non-deployment factors affecting psychological well-being in military personnel: Literature review." *Journal of Mental Health*, 27, no. 1 (2017): 1–11, https://doi.org/10.1080/09638237.2016.1276536
65 Kelley, Britt, Adler and Bliese, *"Perceived organizational support, posttraumatic stress disorder symptoms, and stigma in soldiers returning from combat"*, 2013.
66 Megan Thompson and Donald McCreary, "Enhancing mental readiness in military personnel." In Amy Adler, Carl Andrew Castro, and Thomas Britt (Eds.), *Operational stress. Military life: The psychology of serving in peace and combat: Operational stress*, pp. 54–79 (2006). Westport, CT: Praeger Security International.
67 Paul Bliese and Carl Castro, "The Soldier Adaptation Model (SAM): Applications to peacekeeping research." In Thomas Britt and Amy Adler (Eds.), *Psychological dimensions to war and peace. The psychology of the peacekeeper: Lessons from the field*, pp. 185–203 (2003). Westport, CT, US: Praeger Publishers/Greenwood Publishing Group.
68 Laura Riolli, Victor Savicki, and Everett Spain, "Positive emotions in traumatic conditions: Mediation of appraisal and mood for military personnel." *Military Psychology*, 22, no. 2 (2010): 207–223, https://doi.org/10.1080/08995601003638975
69 Konstantinos Karampas, Michael Galanakis, and Anastasios Stalikas, "Positive emotions, resilience and psychosomatic health: Focus on Hellenic Army NCO cadets." *Psychology*, 7, no. 13 (2016): 1727–1740, https://doi.org/10.4236/psych.2016.713162
70 Donna Harrington, Nadine Bean, Denise Pintello, Deborah Mathews, "Job satisfaction and burnout: Predictors of intentions to leave a job in a military setting," *Administration in Social Work*, 25, no. 3 (2001): 1–16, https://doi.org/10.1300/J147v25n03_01
71 Sylvie Boermans, Wim Kamphuis, Roos Delahaij, Coen van den Berg, and Martin Euwema, "Team spirit makes the difference: The interactive effects of team work engagement and organizational constraints during a military operation on psychological outcomes afterwards." *Stress and Health: Journal of the International Society for the Investigation of Stress*, 30, no. 5 (2014): 386–396, https://doi.org/10.1002/smi.2621

72 Ahmadi Khodabkhsh and Alireza Kolivand, "Stress and job satisfaction among air force military pilots," *Journal of Social Sciences*, 3, no. 3 (2007): 159–163, https://doi.org/10.3844/jssp.2007.159.163.
73 Ibid.
74 Sohyun Han, Frank Castro, Lewina Lee, Meredith Charney, Brian Marx, Kevin Brailey, Susan Proctor, and Jennifer Vasterling, "Military unit support, post deployment social support, and PTSD symptoms among active duty and National Guard soldiers deployed to Iraq," *Journal of Anxiety Disorders*, 28, no. 5 (2014): 446–453, https://doi.org/10.1016/j.janxdis.2014.04.004
75 Yinyin Zang, Thea Gallagher, Carmen McLean, Hallie Tannahill, Jeffrey Yarvis, and Edna Foa, "The impact of social support, unit cohesion, and trait resilience on PTSD in treatment-seeking military personnel with PTSD: The role of posttraumatic cognitions." *Journal of Psychiatric Research*, 86 (2017): 18–25, https://doi.org/10.1016/j.jpsychires.2016.11.005
76 Amy Adler and Carl Castro, "An occupational mental health model for the military." *Military Behavioral Health*, 1 (2013): 1–11, https://doi.org/10.1080/21635781.2012.721063
77 Michael Waller, Susan Treloar, Malcolm Sim, Alexander McFarlane, Annabel McGuire, Jonathan Bleier, and Annette Dobson, "Traumatic events, other operational stressors and physical and mental health reported by Australian Defence Force personnel following peacekeeping and war-like deployments." *Biomed Central Psychiatry*, 12, no. 88 (2012): 1–11, https://doi.org/10.1186/1471-244X-12-88
78 Grace Yan, Lisa McAndrew, Elizabeth D'Andrea, Gudrun Lange, Susan Santos, Charles Engel, and Karen Quigley, "Self-reported stressors of National Guard women veterans before and after deployment: The relevance of interpersonal relationships," *Journal of General Internal Medicine*, 28 (2013): 549–555, https://doi.org/10.1007/s11606-012-2247-6
79 Karen Brouneus, "On return from peacekeeping: A review of current research on psychological well-being in military personnel returning from operational deployment." *Journal of Military and Veterans Health*, 22, no. 1 (2014): 24–29. Retrieved from https://search.informit.com.au/documentSummary;dn=265920949431638;res=IELHEA
80 Michaela Schok, Rolf Kleber, and Hennie Boeije, "Men with a mission: Veterans' meaning of peacekeeping in Cambodia." *Journal of Loss and Trauma*, 15, no. 4 (2010): 279–303, https://doi.org/10.1080/15325020903381873
81 Ibid.
82 Craig Bryan, William Elder, Mary McNaughton-Cassill, Augustine Osman, Ann Marie Hernandez, and Sybil Allison, "Meaning in life, emotional distress, suicidal ideation, and life functioning in an active duty military sample." *The Journal of Positive Psychology*, 8, no. 5 (2013): 444–452, https://doi.org/10.1080/17439760.2013.823557

83 Désirée Stocker, Nicola Jacobshagen, Norbert Semmer, and Hubert Annen, "Appreciation at work in the Swiss Armed Forces." *Swiss Journal of Psychology*, 69 (2010): 117–124, https://doi.org/10.1024/1421-0185.a000013
84 Ibid.
85 Martin Seligman, *Flourish: A new understanding of happiness and well-being—and how to achieve them.*

Valerie Wood and Lobna Chérif

Appreciating the Strengths of Comrades: A Positive Psychology Intervention for Improved Military Leadership, Unit Cohesion, and Member Well-Being

"Leaders don't look for recognition from others, leaders look for others to recognize." – Simon Sinek[1]

Character Strengths and Well-Being

In the early 2000s, the field of psychology had become proficient in the understanding, treatment, and prevention of psychological disorders due to the validation and use of the Diagnostic and Statistical Manual of Mental Disorders (DSM).[2] Noticing this focus on ailment, as opposed to flourishment, Seligman called for a shift towards a more "positive psychology," or a focus on understanding psychological well-being and 'things that are right' with people. Critical to this movement towards a positive psychology were Peterson and Seligman,[3] who created a classification system identifying 24 character strengths (called Values in Action or VIA) that can be broadly organized into six virtues. These virtues have also been empirically validated.[4] The first virtue is *wisdom*, which captures cognitive abilities such as the acquisition and use of knowledge for good purposes, and includes the strengths of creativity, curiosity, judgment, love of learning, and perspective. The second virtue is *courage*, reflecting one's disposition to perform the right act in the face of external or internal opposition, despite resistance and a high risk of loss. Courage is comprised of the more specific strengths of bravery, perseverance, honesty, and zest. The third virtue is *humanity*, which reflects interpersonal strengths such as attending to and befriending others and taking part in acts of generosity and kindness that inspire others. Related character strengths include love, kindness, and social intelligence. The fourth virtue, *justice*, captures civic strengths that underlie healthy community life and accentuate a sense of fairness between people and their larger society. Character strengths included in this virtue are teamwork, fairness, and leadership. The fifth virtue is *temperance*, which consists of having control over excess and having the strengths that protect against it. This virtue is demonstrated through the character strengths of forgiveness, humility, prudence, and self-regulation. Finally, *transcendence* is the sixth virtue, which refers to the extent to which one reflects on life's meaning, and that one is connected to the larger universe. Related character strengths are appreciation of beauty, gratitude, hope, humor, and spirituality.

The strengths, like personality traits, are assumed to be continuous whereby individuals possess each of these strengths to varying degrees, with individuals having a more limited set of core 'signature strengths' which are thought to *most strongly* motivate and guide our behavior. While these signature strengths are strongly expressed and come quite naturally to us, other strengths may be dormant, not having received any deliberate attention from us over the years. There are an almost infinite number of character strengths profiles,[5] and these profiles can change over time. Character strengths are both about "being" (i.e., a part of our identity) but also "doing" in the sense that they involve taking action. In addition, individuals have different ways of expressing each strength. For example, kindness might be expressed through empathy or thoughtful actions, leadership through clear communication or inclusion of multiple perspectives, and gratitude through reflection, or actual expressions of appreciation. We all have a distinctive way of experiencing and expressing our character strengths, and they can also be expressed differently depending on the social context. For example, one individual might strongly express humor and kindness when with friends, love with their family, and perspective and judgment while at work. Individuals can assess their own character strengths profile by completing the VIA Adult Survey and obtaining scores for each of the 24 strengths.[6] While high scores are more reflective of one's signature strengths, low strength scores do not necessarily reflect personal weaknesses. These lower scores simply reflect that there are other character strengths that more strongly reflect one's character. Indeed, it is important to remember that each of us possesses these strengths to *some* extent, and that strengths less central to our character are simply waiting to be activated and developed with the right attention and motivation.

It is important to develop character strengths as they have empirically been shown to contribute to a positive life.[7] Determining what a positive or *good life* looks like or how it is defined has been the source of much debate in the field of positive psychology. However, there does appear to be consensus among researchers that well-being is reflected by both 1) the attainment of pleasure and avoidance of pain (e.g., reporting more positive emotions, fewer negative emotions, more happiness, and life satisfaction), and 2) aspects of life that relate to meaning and purpose including the presence of meaningful relationships.[8] Seligman[9] developed one of the more recent models of well-being that includes the presence of 5 core elements: positive emotions, engagement, positive relationships, meaning, and accomplishment (PERMA). Research has consistently demonstrated a link between the presence of character strengths and indicators of improved well-being including fewer negative emotions, reporting more control over aspects of one's life, experiencing personal growth, greater purpose in life, self-acceptance, and the presence of positive relationships.[10] Wagner and colleagues[11] studied the relationships among all character strengths and the PERMA dimensions and found that while all strengths were positively associated with all PERMA dimensions, there were differences in the magnitude of these relationships. For example, the PERMA dimension *accomplishment* showed the strongest associations with the strengths of perspective, persistence, and zest, while the PERMA dimension *positive relationships* was most strongly associated with the strengths of teamwork, love, and kindness. Overall, this research provides support for the notion that character strengths contribute to well-being.

Character Strengths in the Military

Matthews[12] has argued that the military is "the perfect 'home' for positive psychology", given that the military is comprised of individuals who have been selected based on the absence of pathology, and that the military can be construed as a positive institution, given its strong emphasis on character development, morale, and welfare. In other words, positive psychology principles (e.g., a focus on human flourishing and developing positive qualities) are relevant in the military context. While several military doctrines emphasize the importance of character for successful military leadership (e.g., United States[13], New Zealand[14], Australia[15], Argentina[16], and Canada[17]), there is limited empirical research evaluating character strengths in military populations. Matthews and colleagues[18] investigated whether there were differences in the character strengths of U.S. and Norwegian military cadets relative to an age-matched sample of U.S. civilians, and found that the two military samples were more similar compared to the civilian sample, and were higher on the strengths of honesty, hope, bravery, and teamwork. In a study of applicants to the Australian Army Special Forces, applicants with teamwork as a core strength were 2.6 times more likely to pass the selection process.[19] Boe and Bang[20] found that leadership, persistence, bravery, teamwork, fairness, social intelligence, love of learning, perspective, self-regulation, and creativity were particularly important for succeeding as an officer in the Norwegian Army. More recently, Chérif and colleagues[21] investigated the relationships among military cadet's core strengths and resilience and found that resilience was associated with higher rankings of perseverance, bravery, and humor in their list of signature strengths. This research suggests that the presence of character strengths is important for the well-being of future military leaders.

Strengths-Based Interventions

According to Lyubomirsky and colleagues,[22] there are three main ways that individuals can improve their well-being: a genetically-based set point (i.e., genetic predisposition), life circumstances (e.g., income, winning a lottery), and intentional activities and practices. Strengths-based interventions fall into that third category and involve helping individuals focus on particular strengths and encourage their usage in a new way. These interventions have been identified as highly effective in increasing well-being and ameliorating depression. Indeed, Schutte and Malouff[23] conducted a meta-analysis on the impact of character-strengths interventions on various well-being outcomes and reported that character strengths focused on leveraging one's signature (top five) strengths had a significant impact on positive emotions and happiness, concluding that signature strength interventions have the potential to contribute to beneficial outcomes in various areas of life. Much of this research was based on the use of the signature strengths 'using strengths in new ways' paradigm by Seligman and colleagues[24] where participants receive individualized feedback on their top five character strengths and are instructed to use one of their top five strengths in a new way every day for one week. Therefore, much of the research to date on the effectiveness of character strengths for our well-being has focused on identifying, acknowledging, and improving the use of *one's own* strengths. However, recent research has shown how others' recognition of our strengths (known as 'strengths spotting') can be beneficial.

Recently, Kashdan and colleagues[25] found that romantic couples who reported greater recognition and appreciation of one another's character strengths reported higher levels of indices of relationship quality, but also autonomy (a marker of intrapersonal well-being), relative to those who reported less strengths recognition and appreciation. In addition, Quinlan and colleagues[26] investigated the role of teacher's strength spotting in a strengths-based intervention for elementary students. These researchers found that for students who participated in a character strengths intervention, their teachers were more likely to spot their strengths, and these students also experienced changes in indicators of well-being, and classroom behavior. More importantly, teachers' strength spotting *explained why* those who participated in a character strengths intervention experienced more positive affect, classroom engagement, and need satisfaction (autonomy, competence and relatedness). These findings are consistent with the view that one's own sense of self-worth is based on the extent to which we think others value and accept us.[27] The implications here, are that while interventions geared towards identifying and enhancing our own character strengths have been shown to be quite effective at improving markers of intrapersonal well-being, these effects could potentially be enhanced if the identification and acknowledgment of our strengths came from others. Such interventions might be particularly relevant for military populations, with the quality of one's organizational relationships critical for operational effectiveness and safety. Indeed, if applied to comrade-comrade and leader-subordinate relationships, such interventions might have the potential to increase unit cohesion and perceptions of leadership, both factors which are related to indicators of intrapersonal well-being such as resilience, subjective well-being, and lower prevalence of psychological disorders.[28][29][30]

Proposed Intervention

Initial research suggests that the presence of character strengths might be relevant in predicting success and well-being for some military populations (e.g., recruits, Naval/Officer Cadets). In addition, strengths-based interventions have shown effective at improving well-being, and some promising preliminary findings indicating that spotting and appreciating the strengths of others is associated with indicators of relationship quality and intrapersonal well-being. We propose that strengths-spotting/appreciation interventions might be effective for military populations, where the importance of social cohesion is emphasized both horizontally and vertically across chains of command. Specifically, we propose that interventions aimed at helping comrades and leaders identify, demonstrate, and then express one's appreciation of one another's character strengths could be particularly effective in promoting individual well-being, organizational relationships, and ultimately operational effectiveness.

In-Person Intervention

The following intervention is adapted from the Character Strengths Appreciation intervention developed by Niemiec.[31] Individuals should be assigned to meaningful pairs (comrades that work closely with one another, or immediate supervisor-subordinate pairs). Then, they both take turns assuming the 'strengths appreciator' role. Specifically, the appreciator will follow these four steps. Step

(1) the appreciator carefully examines the list of 24 character strengths with definitions.[32] Step (2) identify what they believe to be their partner's best three (or top three, or most salient three) character strengths. Step (3) write down a recent admirable incident in which their partner displayed each of these strengths. This is an important step, as it is imperative that the strengths appreciation is perceived as being genuine and sincere.[33][34] In other words, how were they seeing the strength expressed in that individual? What impact did that strengths expression have on others, or on the unit's effectiveness? Step 4) Express appreciation by sharing with their partner what they wrote, explaining the importance and value of that character strength for the relationship, and potentially expanding that dialogue to the value of that strength to the unit or organization more broadly. For example, that strength might make it easy to trust that individual, that strength might be inspiring or uplifting to others, and might make the unit function more effectively (and in what way). Then, the pair switches roles so that the recipient is now the strengths appreciator. A debriefing should take place, such that after the partners have completed the activity, they both share what they found valuable about the experience, and what they will take away as a result. In addition, both partners should discuss how they will continue to apply the lessons learned in interactions, duties, and operations going forward.

Virtual Intervention

Focusing on "what's right with others" and spotting strengths in others can be particularly engaging and energizing in virtual settings. The virtual intervention is inspired by the strength-spotting intervention developed by Seligman and colleagues[35] called "You, at your best" and is comprised of 5 steps. Step (1) Participants in a virtual team meeting are invited to de-identify themselves by changing their names/usernames to the same symbol like an ellipsis (i.e., everyone would change their username/displayed name to "..."). Step (2) Participants carefully examine the list of 24 character strengths with definitions. Step (3) Participants choose a particular team member and identify that individual's top character strengths by recalling a specific situation where that individual demonstrated a strength(s) (e.g., 1–3 strengths). To help recall a particular instance, participants are encouraged to think of times where that individual was at their best. Step (4) Participants then use a private message/chat feature to message the individual with their identified strengths. All participants have been de-identified, so the individual receiving the message with strengths will not know who sent the message. The authors recommend that participants repeat this activity such that they send at least one strength to every member of the team. As a result, all participants will receive anonymous identifications of character strengths. Step (5) It is also recommended to have members engage in a team discussion, where participants are invited to share the strengths that they received and their perceptions of those identified strengths. For example, were you surprised by some of the strengths that others spotted in you? Out of the strengths that you received, which were the most impactful for you and why?

Implications for Improved Military Leadership, Unit Cohesion, and Member Well-Being

Beyond improving intrapersonal well-being, potential implications of such an intervention include improvements in perceptions of military leadership (i.e., by promoting a supportive relationship between leader and subordinates), and cohesion (through facilitating or maintaining bonding, comradery). Research overwhelmingly supports the notion that supporting subordinate relational needs through supportive leadership behavior is associated with subordinate emotional well-being, decreased reported stress, and organizational effectiveness.[36] For example, Graves and Luciano[37] found that meaningful and high-quality leader-member exchanges where the employee (among other things) receives encouragement, support, and feels a sense of mutual respect and trust was associated with greater needs satisfaction (needs for autonomy, relationships, and competence), which in turn was related to increased employee well-being, job satisfaction, and organizational effectiveness. In addition, Griffith[38] found that military subordinates who perceived greater leader emotional support (their superiors made attempts to treat them as a person, were interested in their personal welfare, and what they thought about things) reduced the impact of stress on negative outcomes. In addition to soldiers' experience of supportive unit leadership, cooperative peer relations (e.g., cohesion) built their identification with the unit, lessened the likelihood of soldiers leaving the unit and the Army in general, and enhanced their perceptions of their combat readiness. Therefore, opportunities that enhance or promote high quality, meaningful leader-member exchanges and peer interactions are likely to not only strengthen unit cohesion, and perceptions of immediate leadership, but also promote member well-being and operational effectiveness.

In support of the effectiveness of leader strengths spotting of their subordinates more directly, Key-Roberts[39] argued that leaders' recognition of the skills and abilities of their subordinates can provide a distinct advantage for task and mission completion. In particular, leaders who recognize the unique talents of their subordinates will be more successful at mobilizing individuals within the organization to address various challenges. This author identified specific methods by which Army leaders can support their subordinates' well-being and unit success that emerged from the larger body of research exploring the application of strengths-based leadership in military contexts. Consistent with strengths-based leadership, the evidence-based strategies identified that were associated with subordinate development were identification and utilization of their strengths, provision of individualized feedback (focusing on what went right rather than what went wrong or did not go well), building and maintaining a positive climate, and caring and empowering their subordinates. Niemiec[40] has reiterated the importance of complementary signature strengths among individuals, what he calls 'character strengths interpersonal synergies' where members' character strengths can come together and create a synergy in which the new whole is greater than the sum of the strengths (e.g., creativity and self-regulation). Leaders' knowledge of the impactful ways that (even seemingly unrelated) character strengths of unit members can come together could provide the unit with distinct advantages over those where leaders are less attuned to, and less supportive of, subordinate's unique characteristics.

Finally, strengths spotting can be viewed as fostering both transformational and authentic forms of leadership, both of which are overwhelmingly tied to employee well-being and organizational effectiveness.[41,42] In terms of supporting transformational leadership, strengths spotting can support the expression of a fascinating vision, individual support, and intellectual stimulation for followers. In addition, strengths spotting and appreciation can help followers learn to trust themselves and strengthen the belief that followers can do more than they think that they can. In terms of supporting authentic leadership, strengths spotting and appreciation exercises can help to instill trust, transparency, honesty, and perceptions of morality and ethics in leaders, but perhaps most salient is the building of connectedness and quality leader-subordinate relationships[43].

Conclusion

Utilizing strengths-spotting interventions between members or leader-member groups are likely to have a number of desirable outcomes for the member, the unit, and the organization more broadly. Potential implications of the use of interventions like the one proposed, where leaders (or peers) actively spot and express the appreciation of members' unique character strengths, and celebrate them, include a buffering of the negative impacts of stress, enhanced member emotional well-being and needs satisfaction, unit cohesion, perceptions of supportive immediate leadership, stronger commitment to the unit and larger military organization, and organizational and operational effectiveness.

Notes

1 Simon Sinek, *Twitter Post*, March 24, 2014, https://twitter.com/simonsinek/status/448131256304545792.
2 American Psychiatric Association, *Diagnostic and statistical manual of mental disorders*, 4th ed., text rev (2000).
3 Christopher Peterson and Martin E.P. Seligman, *Character strengths and virtues: A handbook and classification*, Vol. 1 (2004). Oxford University Press.
4 René T. Proyer, Fabian Gander, Sara Wellenzohn, and Willibald Ruch, "Strengths-based positive psychology interventions: A randomized placebo-controlled online trial on long-term effects for a signature strengths-vs. a lesser strengths-intervention." *Frontiers in Psychology*, 6 (2015): 456.
5 Ryan M. Niemiec, *Character strengths interventions: A field guide for practitioners* (2017). Hogrefe Publishing.
6 The VIA Adult Survey. *VIA Institute on Character*, https://www.viacharacter.org/survey/surveys/takesurvey.
7 Sanne Theodora Sophia Ghielen, Marianne van Woerkom, and Maria Christina Meyers, "Promoting positive outcomes through strengths interventions: A literature review." *The Journal of Positive Psychology*, 13, no. 6 (2018): 573–585.
8 Lisa Wagner, Fabian Gander, René T. Proyer, and Willibald Ruch, "Character strengths and PERMA: Investigating the relationships of character strengths with a multidimensional framework of well-being." *Applied Research in Quality of Life*, 15, no. 2 (2020): 307–328.
9 Martin E.P. Seligman, *Flourish* (2011). New York, NY: Free Press.
10 Claudia Harzer, "The eudaimonics of human strengths: The relations between character strengths and well-being." In *Handbook of eudaimonic well-being*, pp. 307–322 (2016). Cham: Springer.
11 Lisa Wagner, Fabian Gander, René T. Proyer, and Willibald Ruch, "Character strengths and PERMA: Investigating the relationships of character strengths with a multidimensional framework of well-being." *Applied Research in Quality of Life*, 15, no. 2 (2020): 307–328.
12 Michael D. Matthews, "Toward a positive military psychology." *Military Psychology*, 20, no. 4 (2008): 291.
13 Michael D. Lundy, *The Army's framework for character development*, p. 2 (2017). West Point: U.S. Army Training and Doctrine Command.
14 New Zealand Defence Doctrine (NZDDP-D) (4th ed.). Lieutenant General T.J. Keating (Ed.) (2017). Wellington: New Zealand Defence Force.

15 Commonwealth of Australia (Australian Army), *Land Warfare Doctrine* LWD 0-2-2 Character, Land Warfare Development Centre, Puckapunyal (2005).
16 Ejército Argentino. *Manual del Ejercicio del Mando [Command Manual]*. MFP-51-13. (N.c.) (1990). Argentina: Autor.
17 Government of Canada. *Strong, Secure, Engaged*: Canada's Defence Policy (2017).
18 Michael D. Matthews, Jarle Eid, Dennis Kelly, Jennifer KS Bailey, and Christopher Peterson, "Character strengths and virtues of developing military leaders: An international comparison." *Military Psychology* (2009).
19 Scott D. Gayton and E. James Kehoe, "Character strengths and hardiness of Australian Army Special Forces applicants." *Military Medicine*, 180, no. 8 (2015): 857–862.
20 Ole Boe and Henning Bang, "The big 12: The most important character strengths for military officers." *Athens Journal of Social Sciences*, 4, no. 2 (2017): 161–174.
21 Lobna Chérif, Valerie Wood, and Meaghan Wilkin, "An investigation of the character strengths and resilience of future military leaders." *Journal of Wellness*, 3, no. 1 (2020): 2.
22 Sonja Lyubomirsky, Kennon M. Sheldon, and David Schkade, "Pursuing happiness: The architecture of sustainable change." *Review of General Psychology*, 9, no. 2 (2005): 111–131.
23 Nicola S. Schutte, and John M. Malouff, "The impact of signature character strengths interventions: A meta-analysis." *Journal of Happiness Studies*, 20, no. 4 (2019): 1179–1196.
24 Martin E.P. Seligman, Tracy A. Steen, Nansook Park, and Christopher Peterson. Positive psychology progress: Empirical validation of interventions. *American Psychologist*, 60, no. 5 (2005): 410.
25 Todd B. Kashdan, Dan V. Blalock, Kevin C. Young, Kyla A. Machell, Samuel S. Monfort, Patrick E. McKnight, and Patty Ferssizidis, "Personality strengths in romantic relationships: Measuring perceptions of benefits and costs and their impact on personal and relational well-being." *Psychological Assessment*, 30, no. 2 (2018): 241.
26 Denise Quinlan, Dianne A. Vella-Brodrick, Andrew Gray, and Nicola Swain, "Teachers matter: Student outcomes following a strengths intervention are mediated by teacher strengths spotting." *Journal of Happiness Studies*, 20, no. 8 (2019): 2507–2523.
27 Mark R. Leary and Roy F. Baumeister, "The nature and function of self-esteem: Sociometer theory." In *Advances in Experimental Social Psychology*, Vol. 32, pp. 1–62 (2000). Academic Press.
28 Kara A. Arnold, Nick Turner, Julian Barling, E. Kevin Kelloway, and Margaret C. McKee, "Transformational leadership and psychological well-being: The

mediating role of meaningful work." *Journal of Occupational Health Psychology*, 12, no. 3 (2007): 193.
29 Thomas W. Britt, and Kalifa K. Oliver, *"Morale and cohesion as contributors to resilience."* (2013).
30 J. Du Preez, J. Sundin, S. Wessely, and N. T. Fear, "Unit cohesion and mental health in the UK armed forces." *Occupational Medicine*, 62, no. 1 (2012): 47–53.
31 Ryan M. Niemiec, *Character strengths interventions: A field guide for practitioners* (2017). Hogrefe Publishing.
32 "The 24 Character Strengths" *VIA Institute on Character*, https://www.viacharacter.org/character-strengths.
33 Arran Caza, Gang Zhang, Lu Wang, and Yuntao Bai, "How do you really feel? Effect of leaders' perceived emotional sincerity on followers' trust." *The Leadership Quarterly*, 26, no. 4 (2015): 518–531.
34 Robert A. Baron, "Attributions and organizational conflict: The mediating role of apparent sincerity." *Organizational Behavior and Human Decision Processes*, 41, no. 1 (1988): 111–127.
35 Martin E.P. Seligman, Tracy A. Steen, Nansook Park, and Christopher Peterson. Positive psychology progress: Empirical validation of interventions. *American Psychologist*, 60, no. 5 (2005): 410.
36 Janne Skakon, Karina Nielsen, Vilhelm Borg, and Jaime Guzman, "Are leaders' well-being, behaviours and style associated with the affective well-being of their employees? A systematic review of three decades of research." *Work & Stress*, 24, no. 2 (2010): 107–139.
37 Laura M. Graves and Margaret M. Luciano, "Self-determination at work: Understanding the role of leader-member exchange." *Motivation and Emotion*, 37, no. 3 (2013): 518–536.
38 James Griffith, "Multilevel analysis of cohesion's relation to stress, well-being, identification, disintegration, and perceived combat readiness." *Military Psychology* 14, no. 3 (2002): 217–239.
39 Melinda Key-Roberts, *Strengths-based leadership theory and development of subordinate leaders* (2014). Fort Belvoir, United States: Army Research Inst for the Behavioral and Social Sciences.
40 Ryan M. Niemiec, *Character strengths interventions: A field guide for practitioners* (2017). Hogrefe Publishing.
41 George C. Banks, Kelly Davis McCauley, William L. Gardner, and Courtney E. Guler, "A meta-analytic review of authentic and transformational leadership: A test for redundancy." *The Leadership Quarterly*, 27, no. 4 (2016): 634–652.
42 Bruce J. Avolio and William L. Gardner, "Authentic leadership development: Getting to the root of positive forms of leadership." *The Leadership Quarterly*, 16, no. 3 (2005): 315–338.

43 Meltem Yavuz, "Transformational leadership and authentic leadership as practical implications of positive organizational psychology." In *Handbook of research on positive organizational behavior for improved workplace performance*, pp. 122–139 (2020). IGI Global.

Mala Agarwal

Mindfulness Tools to Combat Stress in Military Personnel

> *"The basic root of happiness lies in our minds; outer circumstances are nothing more than adverse or favorable."*
>
> Mathieu Ricard[1]

Military and Mindfulness: An Introduction

In order to work effectively as a military force, military members at all levels need to develop and use several skills. For instance, all soldiers must understand the situation that they are in, whether they are in a defensive mode or under attack. Soldiers must be aware of what one is communicating while listening to commands and to others' suggestions to form the best action plan. In active combat, soldiers must be completely focused to monitor all the dangers and challenges around them and keep their mind focused to survive and fulfill the mission. Military men and women must do extraordinary things that lie outside the normal human experience.

We propose that a deeper sense of awareness is required under such circumstances. Soldiers need to be acutely aware of what is going on not only externally but also of what is going on within themselves (i.e., internally) in order to reach a delicate balance between navigating in an unusually rapidly unfolding complex environment while maintaining one's self-discipline. In other words, soldiers need to be mindful. Keeping in mind these constraints, we will first describe mindfulness, review reasons for its growing interest in the military and explain how mindfulness training would benefit military personnel. Next, we will discuss some of the myths about mindfulness held by military personnel and suggest ways in which mindfulness can be introduced in the military. Further we will offer examples and mention some mindfulness training practices in the Indian Armed forces. Then, we will review some evidence supporting the role of mindfulness in the military as a low cost, low tech and effective form of mental training that can promote better outcomes and strengthen performance in highly stressful environments. Lastly, we will contemplate future directions for mindfulness implementation in the military.

What Is Mindfulness?

Originally, the concept of what we know as mindfulness emerged in ancient spiritual traditions. Mindfulness is most systematically articulated and emphasized in Buddhism, a spiritual tradition at least 2500 years old. The term mindfulness comes from the Sanskrit word *Smriti*, which literally means and translates to 'that which is remembered'.[2]

The word 'mindfulness' may describe a psychological trait, the practice of mindfulness, a mode or state of awareness, or a psychological process.[3] One of the most cited definitions of mindfulness is the "awareness that arises through paying attention in a particular way on purpose, in the present moment and non-judgmentally".[4] Most other researchers describe mindfulness in a similar way. For example, Ruth Baer defines mindfulness as "the nonjudgmental observation of the ongoing stream of internal and external stimuli as they arise".[5]

Although some researchers focus almost exclusively on the attentional aspects of mindfulness,[6] most follow the model of Bishop and his collaborators, which proposes that mindfulness encompasses two components: (a) the self-regulation of attention, and (b) the adoption of a particular orientation towards one's experiences.[7] Self-regulation of attention refers to the non-judgmental observation and awareness of one's sensations, thoughts or feelings from moment to moment. It requires the ability to switch attention intentionally from one aspect of the external or internal experience to another. The orientation to experience concerns the attitudes that one holds towards one's experience. Mindfulness encourages curiosity, openness and acceptance. Here, acceptance refers to the ability to experience events fully without resorting to either extreme or excessive pre-occupation with, or suppressing thoughts and feelings associated with the experience.

Similarly, Shapiro and her collaborators proposed three main elements of mindfulness,[8] which are intention, attention, and attitude. Intention sets the stage for what is possible, reminding one from moment to moment of why one is practicing mindfulness in the first place. Intention is often dynamic and evolving. For example, highly stressed soldiers may begin a mindfulness practice to reduce hypertension. As their mindfulness practice continues, they may develop the additional intention of relating kindlier to others. The second element is attention. Most people suffer from what is called a 'monkey mind'. One is not really in control over what one is thinking. Thoughts come and go as they please so one might be thinking about something, and then another thought comes up. Mindfulness is just the opposite. It is training one's mind to be present, to stay focused on one simple thing. For instance, it could be focusing on what one is eating, say raisins. After having focused on different aspects of raisins (i.e., taste, texture, etc.), people often say did not know that raisins were so tasty.

Attitude, the third element, refers to the way one pays attention. Mindfulness is about being kind, compassionate and non-judgmental towards oneself. Many individuals tend to judge the situations they find themselves in or their emotions as pleasant or unpleasant, desirable or undesirable, or as good or bad. They have opinions about everything. They struggle to differentiate judgments from facts. Mindfulness is about training to observe our thoughts as thoughts and our emotions as emotions, in a kind and neutral way, without evaluating them. So, in a way, it is about training oneself to be the observer of one's own thoughts, feelings and sensations.

Emerging Role of Mindfulness and Its Benefits to the Military

Soldiers are masters at carrying out the physical activities required by missions, having learned and practiced such skills many times during their training. However, for the most part, their training does not include how to use one's awareness to regulate one's own body and mind during stressful situations. Soldiers do not learn how to pause their mind and relax their body to perceive their environment more clearly and to make better decisions. Also, they lack the necessary skills to align simultaneously their mind and body for action when encountering unexpected difficult hurdles.

When overwhelmed by challenges, being able 'to reset' one's body before encountering the next challenge constitutes an advantage. Unfortunately, for personnel who have not developed this skill, the connection between mind and body may get dysregulated, which in turn may lead to a constellation of reactions that can bring about ineffective behavior. By extension, the way individuals respond to stressful situations has the potential to influence their well-being, happiness, survival, profound sense of interconnectedness, and meaning of life. It all comes down to how one relates to the forces at play in one's life, both external and internal. The message here is that it is not the situation that is problematic, rather it is how one perceives it and responds to it.

In the absence of mindfulness, individuals often react to stressful situations in automatic ways, even before they become aware of what they are doing. A mindful response to challenging situations can influence health and well-being because one is in a better position to make optimal decisions and respond quickly and adaptively to stressors. Because soldiers are the most valuable asset of any military force, it is important to train soldiers' minds to assure military effectiveness. We propose that the most important skill to teach soldiers is to be aware of what is really going on 'within' and 'around' them. This skill can be cultivated through the practice of mindfulness.

There are six reasons that explain why the interest in mindfulness training has been gaining momentum in military circles. First, mindfulness supports soldiers' well-being, which contributes to positive relationships, performance and, ultimately, operational effectiveness. Second, mindfulness is known to be most beneficial to those who suffer from stress,[9] anxiety or depression.[10] Third, addressing mental health issues has become a strategic priority for military organizations. Fourth, military organizations are becoming increasingly concerned about work stress coming from high workloads and view mindfulness as a way to help their personnel deal with numerous work requests. The increasing workloads may erode work-life boundaries, thereby having a detrimental impact on personnel at all levels,[11] especially those who struggle to disengage from work mode.[12] Fifth, mindfulness appeals to military forces because it not only ameliorates health, but it also boosts resilience.[13] As such, mindfulness can be considered a preventive measure,[14] helping personnel in good health to remain healthy. Resilience training may have a short-term impact, but a daily mindfulness practice is more likely to have a sustained effect.[15] Lastly, research has shown that mindfulness can bring a significant improvement in a variety of skills and behaviors highly valued in military settings. Indeed, cognitive functions,[16] sensitivity to others' needs, collaborative working capacity, and the management of complex problems are all ameliorated by mindfulness.[17]

Myths and Ways to Encourage Mindfulness in Military Organizations

In the military, there are many myths about mindfulness. Although mindfulness meditation can help one lead a healthier and happier life, many soldiers resist the idea of practicing it. They fear that taking the time to check mentally on their thoughts, emotions, and sensations might slow them down at critical moments; many think that the practice of mindfulness is a waste of time, at least in part because it is inherently passive.[18] That is, soldiers fear that if they practiced mindfulness, they would lose the edge or they would not be as strategic because they believe that sitting in silence would make their mind passive. On the contrary, sitting in awareness free from judgments and distractions leads to enhanced understanding.

Some soldiers fear that the practice of mindfulness would make one softer and more indecisive.[19] But the opposite is true. Mindfulness facilitates equanimity, calmness and clear and rational thinking. It allows individuals to deal effectively with all situations, regardless of whether tactful or aggressive responses are required. So, mindfulness is very important, especially in combat situations in which important life-threatening decisions must be made swiftly and impulsively, with a clear conscience.

Another common belief among soldiers is that the best way to cope with stressful situations is to repress feelings and to be strong.[20] However, such a coping mechanism can cloud one's options and impair decision making. In fact, the more one tries to appear strong, the less capable one is of thinking flexibly. Practicing mindfulness allows one to move through, giving one the ability to reset and concentrate on one's experience more healthily, fully and positively.

Soldiers may equate mindfulness with relaxation, escape from reality and/or associate it with religious practice.[21] However, mindfulness differs from relaxation because it is about examining what is going on in the mind, body and heart in an ever-changing landscape. That landscape can change rapidly from being relaxed to agitated to distracted. Mindfulness brings awareness to the present moment experience, without repressing anything or trying to make something happen.

Mindfulness is not an escape from reality. It is just the opposite, because mindfulness makes one tune to what is happening in the present moment in a very clear and lucid way, not just in one's external environment but also in one's internal environment.

It is not a religious practice as such, although mindfulness can be practiced along with any religion. While many eastern and western religions have themes of mindfulness, today mindfulness can be practiced by anyone for a multitude of reasons, regardless of their religion. Similarly, it can be practiced by people who simply do not adhere to religious beliefs.

Some soldiers may fear that mindfulness makes them overly mellow[22] or that it reduces grit because mindfulness involves a lot of sitting, But, on the contrary, studies show that it increases vitality and reduces personal distress. Further, individuals who reported being more mindful overall scored higher on a grit scale. In addition, being non-judgmental of one's experiences, thoughts and emotions is associated with greater perseverance.[23]

There are challenges with implementing a mindfulness practice in the military even though the usefulness of mindfulness is increasingly acknowledged by several military forces across the world.[24] For instance, not every mindfulness technique will be appropriate for everyone. Alison Carter and her collaborators[25] warn that mindfulness trainers without formal mental health training may do more harm than good. Indeed, novices practicing mindfulness without training from a practitioner who has formal mental health training may have unpleasant experiences. Sitting in silence for 10–20 minutes or longer may unearth latent trauma in unexpected ways in novices and military population may be particularly vulnerable in this regard. However, eminent researchers have demonstrated that regular mindfulness practice produces actual changes in neural pathways in the brain and that it increases brain functioning and mental endurance.[26]

Thus, because the overall benefits of mindfulness training outweigh the challenges, we propose that mindfulness practice should be introduced in military settings by keeping the following points in mind. The first step is to be clear about the purpose of introducing mindfulness training.[27] It is also important to be clear about its benefits and about what the organization would gain from having its personnel trained in mindfulness. It is also necessary for senior leaders and top management to take ownership of the initiative as they can encourage followers and recruit internal champions. In other words, these initially chosen military members can be 'visible', endorse the mindfulness training and motivate their peers. Later, these members may host and reinforce regular mindfulness sessions after the initial formal training period is over.

Patience is required when introducing mindfulness as there is no immediate discernable benefit. To avoid any confusion, soldiers should be informed of what mindfulness is and what it is not as well as its benefits. They should be encouraged to share which techniques work for them with other members who may struggle with the notion of being mindful.[28]

Military mindfulness programs are still in their infancy. At the time of writing this chapter, there is no unified approach. However, military forces are realizing that mindfulness techniques and programs should be implemented, as they would help team members to strike a balance between personal and team performance.[29] Mindfulness also helps to develop and cultivate creative solutions to work challenges and helps one understand one's role in the team to achieve the mission.[30]

Developing a mindful military organization includes providing educational tools and resources. The organization should offer workshops for military members. Developing mindfulness in an organization does not necessarily mean a focus on formal meditation.[31] Some members may find formal mindfulness practices intimidating, but informal mindfulness practices carried out within daily activities can help members to focus more and be more mindful. It is also not enough to provide resources to develop a mindful organization without an active intervention. Mindfulness is not an activity but a mindset. "It is less about what you do at work than how you do it."[32] Internal communication is important to introduce and promote mindfulness and to encourage military members to adopt its practice.[33]

The implementation of mindfulness training in military forces should begin with leaders because those with a mindful leadership style will foster a people-focused and healthy organizational culture that, in turn, will increase performance. Therefore, mindfulness training should be included in leadership programs, where it is considered part of senior leaders' professional development.[34]

Training programs should encourage members to practice the mindfulness techniques they have learned[35] because the more it is done daily, the more effectively and quickly one can use the techniques at times of stress. Refresher training courses should be incorporated over time. Workplace modifications may be beneficial to better enable members to implement mindfulness practices, such as the inclusion of a space for quiet reflections without distraction. Under ordinary circumstances, some members may experience anxiety or other unpleasant feelings during mindfulness practices.[36] It is therefore suggested that tester sessions[37] should be incorporated and conducted by experienced and skilled professionals. It is also recommended that personnel with mental health issues talk to a mental health professional before starting the practice.

Mindfulness training can be done at the team level.[38] Team mindfulness includes collective mindfulness exercises to understand why team members behave in certain ways. United, a team can reduce the cognitive complexity involved in dealing with stressful situations because the teamwork itself remains psychologically safe, independently of the threat that the team faces.

Examples of Mindfulness Training

Mindfulness Based Stress Reduction (MBSR)

MBSR is a group program developed by Jon Kabat Zinn in the 1970s to treat people struggling with life difficulties.[39] It has since been used effectively with a wide range of population from all walks of life.[40] MBSR is a flexible and customized approach to stress reduction. This 8-week program includes two main components: Mindfulness meditation and yoga. It teaches mindfulness practices that individuals can adapt to best suit their needs.[41] Given the benefits of MBSR, one could hardly argue against offering MBSR, especially because the program does not require an unreasonable amount of time, energy and resources. MBSR, or parts of it, is widely recognized and offered to military populations.[42]

MBSR teaches six principles.[43] First, it encourages approaching the experience of the mindfulness exercise as a challenge rather than as a chore. It is like turning the mindful observation of one's life into an adventure in living. Second, MBSR emphasizes the importance of individual commitment and motivation as well as a regular disciplined practice of meditation, regardless of its form. Third, one must be disciplined as MBSR requires a mindfulness practice for at least six days a week, for a minimum of 45 minutes, for 8 weeks. Fourth, it is important to make each moment count by living it with conscious awareness. Fifth, MBSR adopts an educational rather than a therapeutic approach. Lastly, MBSR has helped in the treatment of various mental illnesses or conditions, such as anxiety, fatigue, grief, pain, and PTSD.

MBSR teaches three useful mindfulness techniques that are worthy of mention. First, the practice known as focus mindfulness emphases a focus inward and the observation of what is happening in one's mind. It can be described as "keeping one's eyes on the road" as it focuses uniquely on the current experience. Awareness mindfulness is the second technique. Here, one must focus on an external instead of an internal awareness, as if one was looking at one's thoughts and feelings from an external perspective, outside of one's usual self-centered perspective, and this, without judgment. The third technique refers to shifting from focus to awareness. It is about observing one's stream of thoughts dispassionately.

MSBR encourages the practice of six popular mindfulness exercises:[44] focusing strictly on the flow of air of each breath; moving one's awareness through one's body, one area at a time, and breathing into each area until it relaxes (body scan); focusing all senses on one object of interest (object meditation); focusing all senses on eating (mindful eating); focusing on body sensations as one is walking (walking meditation); and being mindful of body sensations while stretching (mindful stretching). Other popular awareness exercises include observing one's thoughts and feelings as if they were clouds in the sky or as if they were surfing on a wave (urge surfing).

Several studies have shown that the practice of yoga can benefit people from all walks of life. Yoga can be an excellent way to reduce stress and to practice mindfulness. It can help personnel enhance their levels of calmness, relaxation, and cheerfulness as well as decrease their cognitive and body stress.[45]

Lastly, MBSR teaches meditation. Meditation is not like pressing a button that immediately relieves one's stress level. Rather, its effect on stress relief is gradual. When meditating, individuals take responsibility for their mental state and learn to alter their perceptions of and possibly their reactions to their current circumstances to generate more positive outcomes.[46] The regular practice of meditation facilitates awareness of one's thoughts, emotions and how one experiences stress.

Resilience Mindfulness Training

Raj Marriott[47] adapted Jon Kabut Zinn's 8 week MBSR syllabus to focus specially on the needs of military personnel, both in terms of principles and practices.[48] Initially, the training initiates participants to mindfulness: what it is, how it works, and how it can be helpful to them.[49] The training informs them about the importance of practicing mindfulness daily and discusses ways to use it to maximize performance and to cope with life challenges.[50]

The training content is divided into eight modules:

Module 1 focuses on reducing stress and improving focus. It offers targeted mindfulness exercises to address specific issues. Such exercises can generate significant stress reduction when practiced for even a few minutes daily. It allows participants to address specific stressors, thereby dramatically reducing stress and getting them 'in the right space' to tackle challenges. These exercises can be done with eyes open, while sitting at a desk. One exercise, called the "Reconnect," takes only a few seconds and helps to reduce stress while maintaining a clear focus on what matters most.

Module 2 focuses on cultivating a clear and more sustained focus by letting go of the train of thoughts that generate stress and anxiety. It helps to build intentionality, the ability to stay focused at will and reduces reactions to distractions. Participants learn the skills needed to build their 'mental muscles' to maintain focus and improve resilience

Module 3 helps to develop positive emotions, which helps reduce the distressing thoughts and emotions associated with stress. It helps to overcome the brain negativity bias. Participants become happier and more positively engaged and develop more positive relationships with colleagues.

Module 4 takes the clear focus skill and positive emotions acquired in the previous modules and expands their use to increasingly challenging situations. Specifically, it examines how to cope with challenges and change and how to improve emotional intelligence.

Modules 5 and 6 discuss the key mindfulness skill of creating more mental space between subjective experience and external events. This further helps in reducing the intensity of stress reactions that may lead to unwise decisions and/or procrastination. The purpose of these exercises is to develop a greater presence of mind and clarity of purpose, such that participants can respond quickly, in ways that are more sustainable, more effective and more fully aligned with organizational goals, purpose and norms.

In Module 7, participants learn different ways to apply mindfulness directly to their work, be it writing reports or attending meetings. Techniques of effective communication are emphasized.

Lastly, Module 8 focuses on further establishing and embedding a mindfulness practice into one's life to ensure its benefits continue in the future.

Mindfulness Training in the Indian Armed Forces

Yoga has been practiced in India for over 5000 years. However, it is only recently that the Indian Armed Forces have recognized it as one of the most effective and multi- beneficial tools to combat stress and build resilience among service members. As suggested by S. U. Ray, a Research Officer at the Defense Research Institute in Delhi, India, yoga has its application in the highly stressful environment in all branches of the armed forces (i.e. Army, Navy and Air Force), paramilitary force and police personnel.[51]

These services and voluntary organizations have taken various initiatives to introduce yoga in their wellness and resiliency programs. Such measures are taken to ensure that members of the armed forces work in a healthy environment. Service members and their families practice yoga at military establishments either in person, or virtually at training centers (such as the Army Institute of Physical Training) and yoga institutes.[52] Anecdotes about a service member or a family member taking the initiative to teach yoga at military establishments are common, particularity in the army.

Yoga, as understood and practiced by members of the Armed Forces in India, is primarily Hatha Yoga, a Sanskrit phrase that includes asanas. Asanas refer to steady and comfortable postures or, more specifically, physical poses accompanied by breath work focusing on inhalation and exhalation, thereby moving energy mindfully and effectively throughout one's body to achieve a specific goal. Systematic relaxation, in its simplest form, consists of a body scan from head to toes, mindfully paying attention to each part of the body.[53] A short 3 to 10 minutes systematic relaxation is useful after any vigorous physical exercise, daily physical training (PT) or a hatha yoga practice. Meditation, on the other hand, describes a seated practice done primarily for stilling the mind and that relies on Pranayama (breath work) as a foundation.[54]

Military establishments across the services have contributed also to the creation of a soldiers' resilience program that uses heart-fullness meditation, a technique that includes meditation, cleaning, prayer and relaxation. These tools have been effective to ease stress and anxiety and other imbalances that interfere with a healthy lifestyle.[55]

In 2015, the Indian government has made daily yoga compulsory for all military members, as part of their daily physical exercise.[56] Since then, efforts have been made to educate members and their families about the benefits of practicing yoga. Members are encouraged to download the Ministry of Ayush app –Namaste yoga and its training capsules.[57] The Indian army has also introduced SAHYOG, a program of yoga to reduce stress in the northern and eastern bases.[58]

Yoga has been found to be useful in high altitude, as indicated in a study conducted at the military base in Leh, India. Indeed, yoga helped and facilitated performance and overall health and was found to be superior to routine physical training.[59]

Various training programs are also initiated by voluntary organizations across the nation for the Armed Forces. For example, one such program is a 20-minute mindful version of the Integrated Amrita Meditation Technique (IAM) developed by spiritual master Mata Armritanda Mayi for the armed personnel. It is a simple combination of yoga, meditation, and pranayama. It requires 20 minutes in a day; it is a synthesis of traditional time-tested measures suited for current needs. It has been found to be effective in reducing stress.[60] The Andhra University Department of Yoga and Consciousness offers certificate and diploma courses for members of the defense forces.

Thus, in India, although yoga has not replaced physical training, it has emerged as an effective way to manage stress in the Indian Armed Forces.

Effectiveness of Mindfulness Training in the Military

Pilots / soldiers/ personnel in the military may at times behave in reprehensible ways. Although several factors may contribute to these undesirable behaviors, a dysfunction of the prefrontal cortex is often involved. We, humans, can 'think' and run away. However, when the animal or survival part of our brain takes over, and our cognitive brain shuts off, we may behave in very ineffective ways. Indeed, when our capacity to think has been hi-jacked, our behavior is determined mostly by a combination of fear, anger, and anxiety.[61]

Think, for instance, of an apprentice pilot who, trying to fly an aircraft a little bit higher, in the glide slope, makes an error. When they become aware of their error, their stress reaction kicks in. Emotionally, they may experience anger towards themselves. In this heightened emotional state, their ability to perceive reality may be compromised, the priority being to protect their body. The survival brain will default to the worst-case scenario. They may fail to perceive that the aircraft has derailed from the flight path. The thought that they may be doing this wrong may become so salient that their stress level increases, which in turn may generate further negative emotional reactions such as anxiety. Stress begets emotion, emotion begets stress, and this may spiral to such levels of anxiety and anger that they can no longer think about effective moves they could make to remedy the situation. Instead, they focus their attention on thoughts such as "I am failing" and "I am a failure", thoughts that lead to further emotional reactions. Once the survival brain is activated, the knowledge of what to do in problematic situations while in flight becomes very difficult to access, which may result in ineffective decision-making and behavior.

For this reason, human emotions are important to understand and to monitor. Pilots could benefit from mindfulness training. It is not meant as a substitute for other skills required for flying aircrafts. Returning to the above example, pilots trained in mindfulness would have more readily recognized what their mind was doing after they made a piloting error. They would have more readily recognized that stress and anxiety were normal reactions to what was happening to them at that moment. They would have redirected their attention to the here and now. Such a mental redirection may have been sufficient to interrupt the emotion-stress downward spiral, thereby pausing the survival loop process in the brain. This would have allowed the cognitive brain to focus on the immediate problem and to search for solutions. Hence, the practice of mindfulness can lead to enhanced performance at critical moments.

Although the above scenario is fictional, it was based on a real situation. Indeed, in July 2010, a US Navy pilot was flying her F-18 fighter jet on a combat mission. She had been in the air for 7 hours and was struggling. She recalled all that had been taught to her during her training for eight years to avoid a collision. These hours were brutal, as she mentioned later, and she called them the "worst of her life". Eventually, she remembered the yoga and mindfulness training that she had been practicing while training for deployment. Mindfulness techniques saved her because they helped her to focus on the present moment sufficiently to land safely.[62] Mindfulness meditation had taught her to brush aside harsh self-judgments and allowed her to transform her perspective of the situation from a failure to an opportunity.

In sum, it is important to prepare the military not only for routine work, but also for work in extremely stressful situations because that is the reality in the combat zone.[63] If mindfulness was not taught in training, then military personnel would miss out on a critical skill that could potentially save their life and the lives of others. It has been shown that employing the mindfulness tools has allowed pilots to confront a difficult situation with a positive outlook.[64] In one study, Marines who had been through Mindfulness-based Mind Fitness Training (MMFT) prior to deployment showed improved ability to manage stress.[65] Brian Mockenhaupt noted that soldiers who are calm and focused during chaotic moments are less likely to fire out of fear or frustration.[66] Stanley and her collaborator noted that MMFT builds resiliency and ensures faster recovery from cognitive degradation and psychological injury.[67] Other studies have shown that mindfulness increase well-being, reduces emotional reactivity, and can improve behavior regulation,[68] and soldier resilience. In addition, with respect to breathing exercises, the pranayama practice is known to reduce blood pressure.[69] In sum, the tools of mindfulness have a multitude of benefits for overall well-being and resilience.

What Is Next for Mindfulness?

Mindfulness continues to gain ground in military organizations and in society. The gradually increasing number of studies on mindfulness is expanding the evidence base of its effectiveness. Moreover, with its growing credibility, mindfulness is becoming an acceptable component of military well-being programs.

Research on the impact of mindfulness training specifically in the military is still relatively scarce, but it is an area of growing interest to researchers. Outside the military, the impact of mindfulness is not in doubt. It has demonstrated benefits for building resilience and well-being.

Mindfulness may not be for everyone, but many military members would likely be willing to take up the practice, or at least try it, if given the right opportunity. And if, as the studies tell us, a mindfulness practice improves their well-being and happiness, the organization benefits too, provided mindfulness training is offered for the right reasons and in the right way. There is no doubt that mindfulness is on an upward curve. The signs are that it will become a familiar feature of the military well-being landscape.

Conclusion

The mental training of soldiers is just as important for the armed forces as is the discipline taught through physical training. Historically, this mental training has often been accomplished through challenging and tough training scenarios, such as sleep deprivation. The mindfulness literature suggests that training the practice of focusing one's attention will lead to benefits in mental performance under stress, with the added ability to recover quickly from chaos and shock. Mindfulness training has the potential to be adaptable to typical military schedules and to become a part of the current military culture. In an era when retention of soldiers is difficult and consequences of mistakes are dire, improving the well-being of soldiers individually must become a top priority.

Notes

1. Mathieu Ricard, *"Motivational quotes: Mindfulness quotes for work"* 2018. Retrieved from https:/productivity/theory.com:/mindfulness-quotes-work-career
2. Caraline Brazer, "Roots of mindfulness." *European Journal of Counseling and Psychotherapy* (May 2013), http:// Buddhist psychology typed.com
3. Christopher K. Germer, Ronald. D Siegel, and Paul R. Fulton, "The meaning of mindfulness, what is it? what does it matter?" In *Mindfulness and psychotherapy*, 3–27 (2015). New York, NY, US: Guilford Press.
4. Jon Kabat Zinn, *"Wherever you go, there you are; Mindfulness meditation in everyday life, Part 1 bloom of the present moment"*, pp. 1–47 (1994). P.4. New York, NY: Huyperion (Google Scholar).
5. Ruth Baer, "A Mindfulness training as a clinical intervention: A conceptual and empirical review." *Clinical Psychology: Science and Practice*, 10 (2003): 125–148. P. 125.
6. Kirk Warren Brown and Richard M. Ryan, "The benefits of being present: Is mindfulness and its role in psychological well-being". *Journal of Personality and Social Psychology*, 84, no. 4 (2003): 822–848 (Pub Med).
7. Scott R. Bishop, Mark Lav, Shauna Shapiro, Linda Carlson, Nicole D'Anderson, James Carmody, Michael Speca, Gerald Devins, "A proposed definition of clinical psychology." *Science and Practice*, 11(2004): 230–241 (Google Scholar).
8. Shauna L. Shapiro, Linde S. Carlson, John A. Astin, Benedict Freedman, "Mechanism of mindfulness." *Journal of clinical Psychology*, 62, no. 3 (2006): 373–386.
9. Dillon Small, Nancy Duvall and David Wang, *"Review of the literature regarding mindfulness training as psychological prophylaxis for military stress."* ProQuest dissertation publishing, 2015, http://search.proquest.com/docview/17058702.76/
10. Melba Stetz, Heather McDermott, Brumage Philip Holcombe, Raymond Folen and Ivana Steigman, "Psychological distress in the military and mindfulness based training." *International Journal of Psychology Research*, 7, no. 5/6 (September 1, 2012): 471–484.
11. Shewta Sharma, Neeraj Kumar." Study of work like balance in Indian Soldiers." *Journal of Xian University of Architecture and Technology*, 10, no.9 (2020): 552–558.
12. Donna J .Pickening, *"The relationship between work- life conflict/ work life balance and operational effectiveness in the Canadian forces,* "DRDC Toronto TR 2006–243; Defense R& D Canada Toronto, December 2006.
13. Kelly R. Ihme and Peggy Sundstrom, "The mindfulness shield: The effects of mindfulness training on resilience and leadership in military leaders." *Perspective In Psychiatric Care*, 57, no. 2 (August 2020): 675–698.

14. Grace Bullock, "Preventing stress related mental decline in soldiers", *Mindful organization*. Last updated February 16, 2020, http:// mindful.org -com
15. Maryanna Klatt, Beth Stenberg and Anne Marie Dutchmen, "Mindfulness Based Intervention (MBI) for chronically high stress work environment to increase resiliency and work engagement." *Journal of Visualized Experience. JOVE*, no. 10 (July 2015): e52359–e 52359.
16. Richard Chambers, Barbara Chewen Yeele and Nicholas B. Allen, "The impact of intense mindfulness training on attentional control, cognitive style and affect." *Cognitive Therapy and Research*, 32 (2008): 303–322.
17. Emily K. Lindsay, Brian Chin, Carol M. Greco, Shinzen Young Kirk W. Brown, Aidan G.C. Wright, Joshva M. Smyth, Denna Burkett, and J. David Creswell, "How mindfulness training promotes positive emotions: Dismantling acceptance skills training in two randomized controlled trails." *Journal of Personality and Social Psychology*, 115, no.6 (December 2018): 944–973.
18. Patricia Galaczy, *"The mindfulness and military, the military life style."* MFRC, ESQUI MALT, September23, 2020, http://esquimitic.com
19. Ibid.
20. Ibid.
21. Sheri Jacobson, "Ten myths about mindfulness, need to know." *Counseling Mindfulness, Harley Therapy*, last updates April 19, 2016, httpp://www.Harley therapy.co.uk
22. Jill Suttie, "The myth of mindfulness". *Greater Good Magazine, mind and body*. Last updated September 5, 2015, httpp://greater good Berkley edu.t
23. Ibid.
24. Jeremy Adam Smith, Kira M. Newman, Jill Suttie, and Homa Jazaier, "The state of mindfulness Science." *Greater Good Magazine, mind and body*. Last updated December 5, 2017, http://greatergood,Berkeley,edu.t
25. Alison Carter and Jutta Tobias Murtlock, *"Mindfulness in the military: Improving mental fitness in the U.K armed forces, using next generation team mind fullness training."* Report 525, Institute for employment Studies, May 2019.
26. Richard Davidson, Jon Kabat Zinn, Jessica Schumacher, Melissa Rosenkranz, Daniel Muller, Saki Santorelli, Ferris Urbanowski, Anne Harrington, Katherine Bonus, and John F. Sheridan, "Alteration in brain and immune function produced by mindfulness meditation." *Psychomatic Medicines*, 65, no. 4 (July 2003): 564–570.
27. Paul Adam Mudd, *"7 Whys and 7 ways: How to introduce mindfulness in your workforce community."* Last updated August 7, 2017. Retrieved from https:/thriveglobal.com/shomies/7Whys-7Ways-how-to-introduce-mind-fulness-into-your-workforce
28. Ibid.
29. Gustavo Razzetti, *"How to create a culture of mindfulness."* 2018. Retrieved from https:/blog.liberationist.org/how-to-create a culture-of-mindfulness 12ef0905c8F3.

30. Katherine Te Regama, *"Mindfulness ever matter."* joint base- McGurre-Diixlakchurat, public affairs, 26 April, 2016, https:wwwaf-mil.news
31. Ted Dhanik, *"How to create a culture of mindfulness"* 2017 Retrieved from https:/www.entrepeneur/com/article/294940.
32. Gustavo Rozzetti, "In how to create a mindfulness culture," *Stop Working on Autopilot*, 3rd paragraph.
33. Ted Dhanik, *"How to create a culture of mindfulness."*
34. J. Martunana (n-d), *"What is mindful leadership."* Retrieved from http://instituteformindfulleadership.org.definitions/.
35. Gustavo Rozzetti, *"How to create a mindfulness culture".*
36. David Creswell, "Mindful interventions." *Annual Review of Psychology*, 68 (January 2017): 491–516.
37. Paul Adam Mudd, *"7 Whys and 7 ways: How to introduce mindfulness in your workforce community"*. Last updated August 7, 2017. Retrieved from https:/thriveglobal.com/shomies/7Whys-7Ways-how-to-introduce-mind-fulness-into-your-workforce
38. Alison Carter et al.
39. Jon Kabat Zinn (2013). New York, NY: Bantam Dell.
40. Darmender Kumar, Neeraj Navrattan Sharma, "Mindfulness base stress reduction: An overview". In Hooda D. Sharma (Ed.), *Mindfulness risk and resources*, pp. 197–231, (2013). Delhi: Global Vision Publishing House.
41. Center for mindfulness in medicine health care and society (n-d). Retrieved from http://www.unmassemed.edu/cfm
42. Hyahyang Kang, Jeony Bok Seao, In Yaung Hwang, "Mindfulness based stress reduction, for stress anxiety and psychological well-being: Effects in a Republic of Korea Navy Fleet Crew." *Journal of Psychosocial Nursing and Mental Health Service*, 53, no. 11, (November 2020): 48–55.
43. Jon Kabat Zinn, *"Wherever you go there you are mindfulness meditation in everyday life".*
44. Darmender Kumar, Neeraj Navrattan Sharma.
45. Miciyo Nosaka and Hitoshi Okamura, "A single session of an integrated yoga program as a stress management tool. Companion of daily practice and non daily practice of a yoga therapy program". *The Journal of Alternative and complementary Medicine*, 21, no7 (2015): 444–449.
46. Daniel J. Goleman, and Gang E. Schwartz, "Meditation as an intervention in stress activity." *Journal of Consulting and Clinical Psychology.* 44, no. 3 (1976): 456.
47. Raj Marriott, *"Resilience mindfulness details of training modules"*, https://www.briefmindfulness.com.
48. Karen J. Reivich, Martin E.P. Seligman, and Sharon McBride, "Master resilience training in the US Army." *American Psychologist*, 66, no. 1: 25–34.

49 Catherine Moore, *"Resilience training: How to master mental toughness and thrive"*, http://positivepsychology.com
50 Alhad Annat Pawar, "Resilience training for armed forces personnel: Need for an hour." *Journal of Marine Medical Society*, 19, no. 1 (2017): 4–5.
51 S.U.Ray, *"Importance of Yoga in the Armed Forces"*; Souvenir national yoga week, 2007 MINYIT, 70–73
52 Indian Army Institute of Physical Training, *"Boot camps and military fitness program"*, https://bootcamps,militaryfitnessinstitute.com, 2015.
53 Sarah Guglielmi, *"Fundamentals of systematic relaxation"*, Himalayan Institute, https://himalayaninstitue.org/wisdom-library/workshop-fundamentals, March 2018
54 Swami Rama, *"The real meaning of meditation"*, Yoga international, https://yogainternational.com/article/viewthe real-meaning-of-meditation.
55 Sangeeta Saxena, *"Heartfulness meditation for helping armed forces to increase soldier resilience in no war and no peace time"*, ADU Media, 18 October 2020 https://www-aviation-defense-universe.com
56 Vijeta Sagar, "Yoga made must for Armed Forces", *The Indian Express*, 29 June 2015.
57 Arindam Gosh, "Army goals for virtual yoga", *Times of India* (TNN), 20 June 2021.
58 Anando Bhakto as cited in frontline the *Hindu*, 15 February 2021.
59 "Areas of Work" Ministry of Defense. Government of India, accessed April 13, 2018, https://www.drdo.gov.in/drdo/labs/DIPAS/english/indexnew.jsp?pg=areaofwork.jsp;and research on high altitude was documented in: G. Himashree, L. Mohan and Y. Singh, "Yoga practice improves physiological and biochemical status at high altitude: A prospective case –control study." *Alternative Therapy Medicine*, 22, no. 5 (September 2016): 53–59, https://www.ncbi.nlm.nih.gov/pubmed/?term =yoga +(military +OR+veteran)+himashree.
60 Bringing peace to the Indian Armed Forces and Police Units, May 2019 https://embracingthe world .org
61 Tara Brach, *"The neuroscience of mindfulness and fear."* National Institute for the clinical application of behavioral science, http://www.nicabm.com.blog
62 Tucker Lindsay, *"The Good fight, how yoga is being used within the military."* Last updated September 27, 2018, http//yogajournal.com
63 Victoria Davison, *"Bringing mindfulness to the US Armed Forces."* 21 August, 2018, http:..www.mindful.org.com
64 Katherine Teregama, *"Mindfulness ever matter"*.
65 Douglas C. Johnson et al., "Modifying resilience mechanisms in at-risk individuals: A controlled study of mindfulness training in marines preparing for deployment," *American Journal of Psychiatry*, 171, no. 8 (2014): 844–853, https://www.ncbi.nlm.nih.gov/pmc/articles/PMC4458258/.

66 Brian Mockenhaupt, "A state of military mind," *Pacific Standard*, 8 June 2012 https://psmag.com/social-justice/a-state-military-mind-42839.
67 Elizabeth A. Stanley and Amishi P. Jha, "Mind fitness: Improving operational effectiveness and building warrior resilience," *Joint Force Quarterly*, 55 (2009): 150.
68 Shian-Ling Keng, Moria J. Smoski, and Clive J. Robins, "Effects of mindfulness on psychological health: A review of empirical studies," *Clinical Psychology Review*, (August 2011), https://www.ncbi.nlm.nih.gov/pmc/articles/PMC3679190/.
69 Jeniffer Z. Brandani et al., "The hypotensive effect of yoga's breathing exercises: A systematic review," *Complementary Therapies in Clinical Practice*, (2017).

Tyler E. Freeman[1]

Stress Mindset as an Enabler of Soldier Well-being

Whether it is a result of a fast-paced training environment or the danger and chaos of a battlefield, stress is ubiquitous in the life of a solider. Soldiers experience several nuanced stressors during wartime and each stressor presents its own set of risks to soldiers' well-being. As we begin this chapter, it is important for us to distinguish between physiological stressors – stimuli and events that are harmful to the human body – and psychological stressors – stimuli and events that are harmful to the human mind. What generates stress in human tissue is not the same as what is stressful from a psychological standpoint.[2] Though they are not necessarily mutually exclusive, for the purposes of this chapter, we are concerned with the latter.

Before we delve into examining the effects of stressors on soldiers' well-being, it is necessary to define, at least in a broad sense, the term *stress*. While the literature offers many definitions of stress, this chapter adopts the definition of stress as "a process by which certain environmental demands…evoke an appraisal process in which perceived demand exceeds resources and results in undesirable physiological, psychological, behavioral, and/or social outcomes."[3]

Though it has been the subject of study for nearly a century, the concept of stress has been difficult to operationalize, and consequently measure.[4] One reason for the struggle is the difficulty in distinguishing between direct and indirect effects of stress. Direct effects of stress include those evoked by the physical aspects of a task. In contrast, indirect effects include those evoked by psychological factors resulting from the direct effects. Distinguishing which type of effect is present can be, at times, challenging.[5] For example, reducing the amount of time available to complete a task is likely to be both a direct and an indirect stress. The units of time reduced would be a direct stressor (i.e., less time), whereas the psychological impacts of an additional time constraint, such as increased cognitive load, would be an indirect effect. I mention these effects here because the review to follow does not distinguish between indirect and direct effects of stressors; the distinction is beyond the scope of this chapter. With that in mind, I illustrate a few sources of stress that soldiers encounter. Though this is certainly not an exhaustive list, it serves to illustrate the high-stakes nature of the issue and importance of equipping soldiers with a mechanism for mitigating negative effects of stress on their well-being and performance. A positive stress mindset, an implicit belief that stress can enhance one's performance and well-being, is one such mechanism.

Soldier Stressors

Serving in the military stands among the most stressful jobs one can have.[6] Workload, combat exposure, and the regular requirement to make high-stakes decisions in conditions of uncertainty[7] can all contribute to the psychological stress soldiers experience.

Workload

Sizeable workloads and tight timelines contribute to long work hours and lost sleep, which in turn increase the chances of soldier burnout[8] and hinder effective stress management.[9] This represents a vicious cycle wherein the work demands create psychological stress while also depleting the resources (e.g., adequate sleep) for coping with the stress. In turn, this further exacerbates the psychological impacts of stress and hinders soldiers' ability to manage their time and workloads effectively.

In my work as a researcher focused on enhancing military leaders' performance, I often hear soldiers cite high operational tempos and workloads as a major source of stress. Although these are anecdotal accounts, they are corroborated by research findings.[10] In a 2013 survey of active-duty members of the U.S. Army, 21 % of respondents indicated workload stress is a significant problem. Moreover, it would appear the problem is growing as indicated by a gradual but steady upward trend to 28 % in 2016.[11] Recognizing the significance of workload stress to soldiers' readiness and well-being, the Army has continued to track this trend (though more recent findings were not publicly available at the time of this writing).

Combat Exposure

General combat stress combined with the experience of traumatic events in combat, contribute to a decline in mental health among high numbers of soldiers. In 2011, 19.8 % of U.S. soldiers reported experiencing some sort of psychological issue after a combat deployment.[12] One estimate attributed as many as half of combat casualties to battle fatigue and stress reactions.[13] Though that estimate is dated, it may also be an underestimate as military scholars and practitioners anticipate future battlefields will be more stressful than ever.[14]

Decision-Making in Uncertain Conditions

Now, and even more so in the future, military operations will unfold in a dispersed environment wherein soldiers will regularly be required to make decisions with incomplete and imperfect information.[15] Certainly, making decisions with limited information has been a longtime challenge in the context of war, as evidenced by historical accounts of "the fog of war." However, in future conflicts, the fog of war may result from an overabundance instead of from a lack of information.[16] Vast networks of sensors will provide soldiers with large amounts of data across multiple domains and physical spaces and this information must be processed in compressed timeframes.[17] The availability of information to tactical-level warfighters can lead to significant increases in mental workload. Moreover, future combat operations will likely require information to be collected and acted on at lower echelons than current operations necessitate. The implication is that as younger and less experienced soldiers become increasingly responsible for interpreting information and making high-stakes decisions, there is an increased likelihood they will experience additional psychological stress associated with that decision-making responsibility.

Thus, a dearth of information leads to uncertainty and an abundance of information leads to high cognitive load, both of which contribute to psychological stress. The implication is that regardless of whether soldiers deal with information feast or famine, both decision-making conditions create stress that degrades the cognitive processes (e.g., attention, working memory) that are needed to make high-stakes decisions.

Impacts of Soldiers' Stress

The stressors reviewed above represent only a small selection of the myriad stressors that soldiers may encounter. These stressors can have deleterious effects on soldiers' physical health,[18] emotional well-being,[19] performance,[20] and cognitive functioning.[21] Experienced together or independently, mission success can be negatively impacted.

Stress Impacts on Physical Health

In terms of the physical impacts, stress puts individuals at a greater risk of disease[22] and can lead to dysregulation of active processes of adaptation, a phenomenon referred to as allostasis.[23] Allostasis results in greater wear and tear on the body, referred to as allostatic load.[24,25] Allostatic load may be responsible, at least in part, for the observed linkage between chronic stress and physiological maladaptation in adults.[26] In addition, stressful events in childhood can impact individuals' health well into adulthood,[27] which suggests that stress can have cumulative effects over decades that degrade the physical health of adults.[28] It is well established that physical fitness promotes adaptive responses to stress,[29] even among soldiers during grueling events.[30] Recent evidence suggests the experience of stress is inversely related to physical activity,[31] which has led researchers to pose physical activity as a potential moderator of the link between stress and negative health outcomes,[32] such as greater risk of cardiovascular maladies,[33] weakened immune responses,[34] increased likelihood of infections,[35] and a compromised nervous system.[36] Individuals experiencing acute and chronic stressors may also incur increased costs associated with mental healthcare owing to the impacts of stress on emotional well-being.

Stress Impacts on Emotional Well-being

Emotional well-being refers to the emotional quality of individual experiences and positive emotional responses to negative events. The ability to experience positive emotions and moods – as well as the ability to regulate negative emotions and moods – is a critical antecedent of work performance in military organizations.[37] Unfortunately, the stress soldiers experience during deployments brings significant and deleterious effects to their emotional well-being.[38] For example, deployment stress has been associated with high rates of mental disorders and alcohol abuse.[39] Additionally, a 2008 survey indicated 13 % of soldiers returning from Iraq or Afghanistan had recently considered suicide.[40] Deployment stress is also linked with sharp decreases in soldiers' mood[41] and morale.[42] Moreover, emotional well-being has been posited as a critical protective factor that helps mitigate the negative effects of stressors on soldiers' work performance.[43]

Stress Impacts on Work Performance

In 2008, 23 % of American soldiers suggested their work performance suffered because of stress.[44] Of those soldiers, 15 % asserted stress hindered their ability to do their jobs, and 13 % reported this hindrance led to increases in concern from their supervisors.[45] All soldiers work as part of larger teams wherein each soldier's responsibilities and tasks are interdependent on the work of others. The interdependent nature of teamwork creates a high probability that when stress hinders individual performance, it will also hinder team performance. Stress among team members is associated with a decrease in helping behaviors, increased conflict, inattention to social or interpersonal cues, and less cooperation.[46] In sum, stress can impair soldiers' work performance across different contexts through a variety of mechanisms. Because work performance and cognition are largely intertwined, it is reasonable to assert that among those mechanisms are the deleterious impacts stress has on soldiers' cognition.

Stress Impacts on Cognition

In addition to its effects on soldiers' physical and emotional health, stress is associated with impairments to several cognitive functions.[47] Stress reduces attentional focus, influences perceptual processes, increases cognitive load, boosts the demand for cognitive resources such as working memory, degrades encoding and retrieval of short and long-term memories and, stifles creative problem solving.

Attention and Perception. Although attention and perception are distinct processes, they are intrinsically linked, especially in respect to real-world decision making. Perceived stress increases distraction and reduces attentional focus.[48] Given the importance of situation awareness for soldiers' ability to make good decisions, an increase in distraction or a reduction in attentional focus may lead to a decrease in the utilization of relevant environmental cues during the decision-making process.[49] In life-threatening situations, time perception can become distorted,[50] which can alter other aspects of perception, such as the detection of the threats in an operational environment.[51] Thus, in dangerous situations that require soldiers to attend to multiple cues, stress-related decrements to attention and perception can potentially degrade soldiers' situational awareness, the quality of decisions, their performance, and ultimately their survivability during combat.

Working, Short-Term, and Long-Term Memory. Most stress researchers agree that stress reduces working memory capacity. Working memory is a limited cognitive resource that enables information processing and encoding into long-term memory.[52] Originally conceptualized as distinct mechanisms,[53] recent research suggests that information processing and information storage in working memory are not independent functions and that they draw on the same limited resource, namely, attention.[54] Researchers examining the effects of combat-like conditions on the cognitive performance of sailors and soldiers observed severe decrements among participants' attention and working memory.[55] Unsurprisingly, these decrements were associated with impairments to cognitive functions that depend on attention and working memory, such as logical reasoning and short-term memory. Alarmingly, researchers noted that cognitive deficits resulting from combat-like stress were greater than deficits typically produced by alcohol intoxication, sedative drugs, and clinical hypoglycemia.[56]

In general, explanations for how stress impacts short- and long-term memory center on reductions in working memory. For example, stress may cause attentional tunneling, which will limit the ability of soldiers to encode cues and other environmental information into short- and long-term stores. In this regard, the influence of stress on memory has similar limiting or reductive effects as it does with attention. Additionally, neuroscientific research suggests a physiological connection described as humans' biochemical stress response (e.g., cortisol release) that may have a direct negative impact on hippocampal functioning resulting in memory decrements.[57]

Creativity and Problem Solving. Generating alternative courses of action for solving novel problems draws heavily upon soldiers' creative processes and these creative processes are preceded by basic information processing and interpretation of available information. Consequently, deleterious effects on these prerequisite cognitive processes will adversely affect soldiers' ability to solve novel and stressful problems.

When making decisions and completing tasks while under stress individuals consider fewer alternatives.[58] This may be a direct result of, for instance, time constraints or a result of fewer cognitive resources to facilitate a comprehensive search for alternatives. Researchers have observed that under high life stress, individuals who were inclined to attend to internal physiological states considered fewer possible solutions to problems and were more likely to re-examine ineffective alternatives.[59]

Stress Impacts Vary between Individuals

In sum, a large body of evidence supports the assertion that psychological stress has diffuse effects of soldiers' physical health, emotional well-being, work performance and cognitive function. However, evidence also demonstrates that the magnitude and direction of these effects is variable between individuals. Lazarus and Eriksen found stressful conditions did not produce similar and reliable effects across participants in controlled laboratory research.[60] After being subjected to the same stressful conditions, Lazarus and Eriksen observed some people experienced negligible levels of stress whereas for others, the stress was significant. They also found that in stressful conditions, and depending on the task, some people's performance remained unchanged or suffered whereas others' performance improved. What stress researchers soon realized was the need to account for individual differences (e.g., in motivational and cognitive variables) to understand the mediating factors between stressors and the stress response as well as factors that promote a positive stress response which is referred to by some, as "eustress."

Adaptive Stress Responses and the Eustress Concept

In 1974, the concepts of eustress and distress were introduced to differentiate the types of responses to stress.[61] Hans Selye described eustress as a beneficial, adaptive stress response whereas he associated distress with the deleterious impacts of stress.[62] Richard Lazarus discussed eustress as a positive cognitive response to a stressor that is associated with positive feelings and a healthy physical state whereas distress is a severe stress associated with negative feelings and physical impairments.[63] Similar to Lazarus' emphasis on the importance of cognitive mediators of the stress response, researchers have more recently posited that whether the stress experience is eustress or distress is determined by individuals' *perception* of the intensity of the stressor, its source, duration, controllability and desirability.[64]

In their research in the 1960s, Lazarus and his colleagues created stressful conditions by presenting subjects with videos of people experiencing severe injuries.[65] They observed that by simply changing the way participants were psychologically oriented to the films had significant effects on their stress response. The stress response was much less severe among participants who were told the injuries were staged and nobody was hurt in comparison to subjects who were told the people in the videos suffered severe pain and infections. Other research around this time showed the severity of a stress reaction was dependent upon subjects' evaluative thoughts regarding an impending stressor.[66] Ultimately, stress research in the 1960s and 1970s was instrumental in bringing about changes in the way stress researchers at the time were thinking about stress and psychologists began to incorporate human cognition into their explanations of variability in people's responses to stress. This was the beginning of a new zeitgeist in the study of psychological stress that set the stage for Lazarus and Folkman's introduction of cognitive appraisal theory.[67]

Cognitive Appraisal Theory

According to Richard Lazarus, the experience of stress is determined not only by the presence of stressors in the environment but also individuals' responses to those stressors.[68] Lazarus and Folkman, in their conceptualization of cognitive appraisal theory, stated that cognitive appraisal occurs when an individual reflects on: (a) how threatening a stressor is and (b) the resources that are required to remove the stressor and/or assuage the stress it produces.[69] In general, cognitive appraisal consists of two appraisal types, primary and secondary appraisal.

During primary appraisal, individuals assess the demands posed by a stressor and may ask themselves questions such as, "What does this stressor mean?" and "How does it affect me?"[70] Subsequently, individuals typically determine that the stressor is either: (a) unimportant, (b) a positive experience or (c) a negative experience.

Counterintuitively to its name, secondary appraisal occurs in parallel with primary appraisal and involves the emotional and motivational aspects of stress appraisals. During secondary appraisal, individuals assess the resources and ability to meet the demands posed by the stressor. When demands exceed the resources it yields a negative appraisal of the stressor as a threat, which can cause distress. When the resources exceed the demands, it yields a positive appraisal of the stressor as a challenge that can be overcome.[71] A positive appraisal would be indicated by statements like "I will give it my best shot" whereas a negative appraisal and mindset would be indicated by statements like "There is no way I can overcome this."[72] According to cognitive appraisal theory, individuals who appraise stress positively demonstrate better performance outcomes and are more likely to perceive stressors as challenges to be overcome rather than threats.[73]

To put this in context, suppose a soldier in a combat environment is notified there is a strong enemy presence in a particular area. That soldier might appraise the enemy presence simply as part of the expected environment. Or that soldier may appraise the enemy presence as strategically good because that area is full of friendly forces with greater capabilities. Alternatively, the soldier may appraise the enemy presence as stressful because they are currently in the area with only a small force. After this primary appraisal, the secondary appraisal involves the classification of the enemy presence as either a threat that can cause future harm, a challenge that motivates performance, or a harm-loss where the damage is already done. In other words, individuals' perceptions of stressors seem to influence the outcomes of the stress experience and this position is supported by recent research to conceptualize Stress Mindset Theory.

Stress Mindset Theory

"Mindsets" are defined as "*implicit theories* about the malleability of human characteristics" and have been shown to influence important outcomes.[74] For example, the mindset that students hold about the malleability of their intellectual abilities has an impact on academic achievement.[75] Specifically, students who hold the mindset that intellectual ability is fixed are more likely to perceive academic challenges as a sign that they are intellectually unfit, while students whose mindsets support the idea that intellect can be developed are more likely to exert effort toward overcoming academic challenges.[76] A similar effect of mindset has surfaced for mindsets regarding the malleability of social skills.[77]

Applied to stress, *stress mindsets* can be defined as people's implicit beliefs about the nature of stress. More specifically, individuals who believe *stress is enhancing* can be distinguished from those who believe that *stress is debilitating*. The stress-is-enhancing mindset (SEM) reflects the belief that stress enables people to perform their best, whereas the stress-is-debilitating (SDM) mindset reflects the belief that stress is detrimental to people's performance.

Several interventions for mitigating stress effects have focused on reappraisal, avoidance of stressful situations, and psychoeducation on risks and negative consequences of stress. A commonality among these interventions is that stress is represented as something that needs to be managed *because* it is detrimental to performance. However, a growing body of evidence suggests individuals' stress mindset can bring about changes in psychological and physiological states,[78] and researchers have grown interested in understanding whether such mindsets also influence the individuals' perception (e.g., appraisal) of the stressor. Similar to the concept of eustress, previous psychological research indicated stress also has enhancing influences on memory,[79] physical health,[80] social outcomes, and self-perceptions.[81] Thus, researchers hypothesized that SEMs may promote beneficial responses to stress, including promoting better physical health, emotional well-being, cognitive functioning, and work performance than SDMs.

Alia Crum and her colleagues demonstrated that as individuals adopted SEMs they experienced fewer stress-related decrements to their mood and increments to their anxiety compared to individuals with SDMs.[82] Furthermore, the authors observed that as mindsets moved from a SDM toward a SEM, individuals' health and work performance improved. The authors also found SEMs were associated with adaptive physiological responses to stressful events, as measured by changes in baseline cortisol levels.

In later research, Crum and her colleagues found that, relative to SDMs, SEMs were associated with greater positive affect, attentional bias towards happy faces, and more cognitive flexibility.[83] From the vantage point of the cognitive appraisal theory, Kilby and Sherman found that individuals with a SEM were more likely to appraise stressful situations as challenges than threats.[84] Similarly, Casper and colleagues showed that in response to daily workload anticipation, SEMs were associated with greater approach-coping efforts throughout the workday compared to those with SDMs.[85]

In sum, recent research findings suggest that compared to individuals with SDMs, individuals with SEMs are more likely to realize positive health-and performance-related outcomes when faced with stressful events. This is good news for those with SEMs but begs the question of how to help individuals with SDMs. Fortunately, stress mindset research demonstrates stress mindsets are malleable and illustrates the effectiveness of simple interventions for modifying people's implicit beliefs about stress to promote the development of a SEM.[86]

The Stress Mindset Intervention

Alia Crum and her colleagues designed and evaluated a mindset-training program to promote SEMs and found the training program had positive effects on self-reported health, performance, and well-being among employees in a financial institution, in comparison to a control group.[87] Employees reported significant improvements in their overall satisfaction with their health, better performance at work (with respect to collaborating, being engaged, creating new ideas, and sustaining focus), and improvement of subjective physical symptoms. Statistical analyses revealed these improvements occurred through changes in mindset initiated by the training program.

Two years later, Crum and her colleagues reported the methods and results of three studies that extend the findings of their previous research.[88] In the first study, they developed and validated a stress mindset measure (SMM). Using the newly validated SMM in a second study, the authors demonstrated that stress mindsets are (a) malleable via video priming (i.e., less than 10 minutes total of video exposure over the course of a week), (b) that changes in stress mindset are associated with changes in mood and anxiety and, (c) that SEMs are associated with work performance gains. In a third study, Crum and her colleagues investigated how stress mindsets influence physiological markers of stress and found SEMs were related to more adaptive cortisol profiles. Noting that elevated[89] *and* blunted[90] levels of cortisol secretion are associated with increased arousal and have potentially negative health consequences, the authors hypothesized individuals with SEMs would exhibit moderate (i.e., neither elevated nor blunted) levels of cortisol secretion. This hypothesis was confirmed. For those with high cortisol reactivity to stress, having a SEM lowered cortisol secretion in response to a stressful stimulus. In contrast, for those with low cortisol reactivity, a SEM was associated with increased cortisol secretion. In sum, over the course of the three studies, Crum and colleagues demonstrated

(a) stress mindset is a distinct construct from other stress-related variables, (b) that stress mindset can be manipulated with a relatively simple video-based approach and, (c) that mindset influences important physiological responses to stress. These findings provide ample support for stress mindset as an avenue of approach for interventions aimed at promoting soldiers' adaptive and productive responses to psychological stress.

The Impact of Stress Mindsets of Service Members' Performance

Although the evidence shows stress mindsets are malleable and that, relative to SDMs, SEMs are associated with more adaptive stress responses, the evidence emerged through research wherein the participants were recruited from industry or academia. Moreover, the focal stressors in past research were characteristic of the types of stressors civilians may encounter on the job, such as delivering a public speech.[91] However, the severity of the stress derived from fear of a botched public speech pales in comparison to the stress soldiers experience in combat, or even during realistic training. Thus, researchers sought to investigate whether the effects of stress mindsets shown in past research could be found in a military context where service members face stressors of significantly greater magnitude and consequence.

Smith, Young and Crum[92] conducted research with Navy SEALs during the infamously grueling Basic Underwater Demolition/SEAL (BUD/S) training to examine the extent to which various mindsets predicted trainees' performance and success, as well as instructor and peer evaluations of trainees. In addition to stress-is-enhancing mindsets the authors examined the impact of failure-is-enhancing mindsets and non-limited willpower mindsets on metrics of trainees' success. A failure-is-enhancing mindset is defined as the implicit belief that failure promotes learning, growth, and performance. The non-limited willpower mindset refers to the implicit belief that willpower can be maintained or enhanced with effort. The authors reasoned that because the failure-is-enhancing and non-limited-willpower mindsets are conceptually related to the stress-is-enhancing mindset, they may predict trainee outcomes in a similar manner. The authors examined the influence that failure and willpower mindsets have on trainee outcomes to differentiate it from the influence of stress mindsets on individuals' performance and resilience in an extremely stressful context.

Results of the study demonstrated that BUD/S trainees with greater stress-is-enhancing mindsets had improved performance, lasted longer in the training program, had increased rates of program completion, and were more positively evaluated/rated by instructors and peers. In contrast, trainees with a failure-is-enhancing mindset[93] had poorer outcomes such as slower obstacle course performance, higher dropout rates, and did not persist as long in the training in comparison to trainees with stress-is-enhancing mindsets. The authors noted this is inconsistent with findings in educational settings but consistent with findings from evaluative settings. BUD/S is an evaluative setting and moreover, trainee signs of failure can prompt instructors to push a trainee harder to see if they will succumb to, or overcome, the stress and pressure. In other words, failure can lead to increased demands during BUD/S and therefore the authors suggested it is unsurprising a failure-is-enhancing mindset was associated with more negative outcomes. However, this is not to say that a failure-is-enhancing mindset is not beneficial in other military performance contexts.

Similar to findings regarding failure mindsets, trainees with a non-limited willpower mindset[94] received more negative instructor and peer evaluations. The authors suggested this finding may indicate an emergence of negative social consequences for those holding non-limited willpower mindsets. One potential explanation the authors offered is that, owing to the belief that willpower can be enhanced with effort, trainees with non-limited willpower mindsets may have higher demands and less empathy for others experiencing challenges during training. Contrary to popular belief, BUD/S is not a "dog eat dog" training program. Instructors value trainees' demonstration of teamwork, cooperation, and support despite the severely stressful conditions. Thus, if trainees with a non-limited willpower mindset were dismissive of struggling trainees, they may have received more negative instructor peer and evaluations because they were perceived as less supportive of their fellow trainees.

Smith and collaborators concluded that even in an environment with extreme physical and mental stressors, individuals with SEMs exhibited better training outcomes and received more positive evaluations compared to related failure-is-enhancing and non-limited willpower mindsets. Thus, the results of this research suggest stress mindsets may be a more significant enabler of performance in highly demanding and stressful environments. The authors called for future research to investigate the psychological and physiological mechanisms that underlie the relationship between stress mindset and performance.

Conclusion

The study of the role of mindsets in shaping reactions to stress is in a stage of relative infancy. Yet, despite this, a burgeoning body of evidence supports the position that SEMs can help buffer the deleterious impacts of stress on individuals' well-being and performance. Research suggests SEMs have beneficial impacts in highly stressful military contexts. Moreover, the evidence demonstrates stress mindsets are quite malleable, which is critically important for justifying interventions aimed at engendering SEMs among soldiers.

Research also indicates it does not require extensive time or effort to engender SEMs, at least among civilians. My related research experience with U.S. military personnel (which I am not authorized to discuss as of the time of this writing) suggests a similarly simple, video-based stress mindset intervention may effectively engender positive shifts toward SEMs among service members. That a relatively simple, short, video-based stress-mindset intervention can promote SEMs is a significant strength that highlights the potential utility of a SMI in military populations. As discussed above, soldiers often sustain high workloads. Between their daily duties and training requirements, there is little time available to engage soldiers with elective interventions that promote their well-being. Thus, that research has demonstrated such strong and positive effects on physical, health, and performance outcomes with as little as 30 minutes of exposure to stress-mindset interventions suggests that implementing such interventions has significant potential for enhancing and maintaining the overall well-being of service members.

But there remains more work to be done. For example, although research has shown a relationship between stress mindsets and physical performance during training,[95] future research would benefit from more robust measures of physical performance. Biometric techniques (e.g., heart rate variability and V02 Max) would afford researchers objective indicators of soldiers' physiology during performance of physically demanding tasks, such as standardized military fitness tests. Research should be designed to account for likely covariates (e.g., weather conditions, age, body mass) in order to obtain a clear understanding of the precise impact stress mindsets can have on physical performance.

Regarding the robustness of the effect of the SMI, it will be important for future research to examine the decay rate of positive gains in stress mindsets. Although research has demonstrated positive shifts in stress mindset following exposure to a SMI,[96] it is uncertain if, and for how long, those gains would be sustained. Research is needed to determine the robustness of these effects and gauge the frequency with which service members must be re-exposed to a SMI to sustain a SEM.

It is my hope that scientists' interest in understanding the psychological and physiological mechanisms and benefits of stress mindsets continues, and that practitioners pursue work to develop and refine interventions aimed at engendering SEMs among service members. Military service is among the most stressful jobs one can have. It is incumbent upon scientists and practitioners alike, to equip service members with tools that foster their performance and well-being in the stressful conditions soldiers often encounter during their service to their Nation.

Notes

1. I would like to thank Dr. W. Anthony Scroggins for his insightful feedback and invaluable contributions to this chapter and, more broadly, for his dedication to improving service members' ability to remain effective despite the unavoidable stress they experience.
2. Richard S. Lazarus, "From psychological stress to the emotions: A history of changing outlooks." *Annual Review of Psychology*, 44, no. 1 (1993): 1–22.
3. Eduardo Salas, James E. Driskell, and Sandra Hughes, "The study of stress and human performance." In *Stress and human performance*, pp. 6 (1996). Mahwah, NJ, Lawrence Erlbaum Associates, Publishers.
4. Salas et al., "Stress and human performance."; Staal, Mark A. *Stress, cognition, and human performance: A literature review and conceptual framework* (2004). Hanover, MD: NASA.
5. Staal, *Stress* (2004).
6. Simone R. Johnson, "The top 10 most and least stressful jobs." *Business News Daily*, November 27, 2019.
7. Jamie Halchishick, "NCOs, mission command, and their future as part of Multi-Domain Battle" *LinkedIn*, August 2017, Retrieved from: https://www.linkedin.com/pulse/ncos-mission-command-future-part-multi-domain-battle-halchishick/; Peter D. Harms, Dina V. Krasikova, Adam J. Vanhove, Mitchel N. Herian, and Paul B. Lester, "Stress and emotional well-being in military organizations." In *The role of emotion and emotion regulation in job stress and well being* (2013). Emerald Group Publishing Limited.
8. Evangelia Demerouti, Arnold B. Bakker, Friedhelm Nachreiner, and Wilmar B. Schaufeli, "The job demands-resources model of burnout." *Journal of Applied Psychology*, 86, no. 3 (2001): 499–512.
9. William D.S. Kilgore, Ellen T. Kahn-Greene, Erica L. Lipizzi, Rachel A. Newman, Gary H. Kamimori, and Thomas J. Balkin, "Sleep deprivation reduces perceived emotional intelligence and constructive thinking skills." *Sleep Medicine*, 9, no. 5 (2008): 517–526.
10. Ryan P. Riley, Katherine J. Cavanaugh, Ryan L. Jones, and Jon J. Fallesen, "*The 2016 Center for Army Leadership Annual Survey of Army leadership (CASAL): Military leader findings (Technical Report 2017-1).*" Fort Leavenworth, KS: Center for Army Leadership (2017).
11. Riley et al., "CASAL," (2017).

12 Joint-Mental Health Advisory Team 7, "*Joint Mental Health Advisory Team (J-MHAT) 7 Operation Enduring Freedom 2010 Afghanistan.*" Report chartered by the Office of the Surgeon General United States Army Medical Command, Office of the Command Surgeon HQ USCENTCOM, and Office of the Command Surgeon US Forces Afghanistan (USFOR-A), 2011.
13 Thomas R. Mareth and Alan E. Brooker, "Combat stress reaction: A concept in evolution." *Military Medicine*, 150, no. 4 (1985): 186–190.
14 Jon Bott, John Gallagher and Josh Powers, "Multi-domain battle: Tactical implications." *Over The Horizon*, August 28, 2017.
15 Halchishick, "NCOs," (2017).
16 U.S. Department of the Army, *Multi-domain battle: Evolution of combined arms for the 21st century: 2025-2040. Version 1.0* (December, 2017). Washington, DC: Author.
17 David G. Perkins, "Multi-domain battle: Driving change to win in the future." *Military Review* (July–August 2017).
18 C.G. Weiman, "A study of occupational stressor and the incidence of disease risk," *JOEM*, 19 (1977): 119–122; John H. Milsum, "A model of the eustress system for health/illness." *Behavioral Science*, 30, no. 4 (1985): 179–186; Hans Selye, *The physiology and pathology of exposure to stress, a treatise based on the concepts of the general-adaptation syndrome and the diseases of adaptation* (1950). Montreal, Acta Inc. Medical Publishers.
19 Richard Lazarus, "From psychological stress to the emotions: A history of changing outlooks." *Annual Review of Psychology*, 44 (1993): 1–21; Majid Lotfalian, Seyed Fazlollah Emadian, Nahid Riahi Far, Maryam Salimi and Fatemeh Sheikh Moonesi, "Occupational stress impact on mental health status of forest workers." *Middle-East Journal of Scientific Research*, 11, no. 10 (2012): 1361–1365; Shauna L. Shapiro, Daniel E. Shapiro and Gary E.R. Schwartz, "Stress management in medical education: A review of the literature." *Academic Medicine* (2000); Shauna L. Shapiro, Kirk W. Brown, and Gina M. Biegel, "Teaching self-care to caregivers: Effects of mindfulness-based stress reduction on the mental health of therapists in training." *Training and Education in Professional Psychology*, 1, no. 2 (2007): 105, http://dx.doi.org/10.1037/1931-3918.1.2.105
20 Teri Saunders, James E. Driskell, Joan H. Johnston, and Eduardo Salas, "The effect of stress inoculation training on anxiety and performance." *Journal of Occupational Health Psychology*, 1, no. 2 (1996): 170–186; James E. Driskell, Eduardo Salas and Joan H. Johnston, "Decision making and performance under stress." In T. W. Britt, A. Adler, & C. A. Castro (Series & Vol. Eds.) *Military life: The psychology of serving in peace and combat: Vol. 1 military performance*, pp. 128–154 (2006). New York, NY: Praeger Press.
21 Arthur W. Combs and Charles Taylor, "The effect of the perception of mild degrees of threat on performance." *The Journal of Abnormal and*

Social Psychology, 47, no. 2S (1952): 420–424; Francis T. Durso, and Amy L. Alexander, "Managing workload and situation awareness," In Eduardo Salas and Dan Maurino (Eds.), *Human factors in aviation*, pp. 217–247 (2010). Burlington, MA, Elsevier; Staal, *Stress* (2004).

22 Sheldon Cohen, Ronald C. Kessler and Lynn U. Gordon, "Strategies for measuring stress in studies of psychiatric and physical disorders." *Measuring Stress: A Guide for Health and Social Scientists* (1995): 3–26.

23 Matthew A. Stults-Kolehmainen and Rajita Sinha, "The effects of stress on physical activity and exercise." *Sports Medicine*, 44, no. 1 (2014): 81–121, https://doi.org/10.1007/s40279-013-0090-5

24 Bruce S. McEwen, "Physiology and neurobiology of stress and adaptation: Central role of the brain." *Physiological Reviews*, 87, no. 3 (2007): 873–904.

25 Burton Singer and Carol D. Ryff, "Hierarchies of life histories and associated health risks." *Annals of the New York Academy of Sciences*, 896, no. 1 (1999): 96–115.

26 David Borsook, Nasim Maleki, Lino Becerra, and Bruce McEwen, "Understanding migraine through the lens of maladaptive stress responses: A model disease of allostatic load. *Neuron*, 73, no. 2 (2012): 219–234; Stults-Kolehmainen, "Effects of stress," (2014).

27 Andrea Danese and Bruce S. McEwen, "Adverse childhood experiences, allostasis, allostatic load, and age-related disease." *Physiology & Behavior*, 106, no. 1 (2012): 29–39; Peggy A. Thoits, "Stress and health: Major findings and policy implications." *Journal of Health and Social Behavior*, 51, no. 1 (2010): S41–S53.

28 Gregory Miller, Edith Chen, and Steve W. Cole, "Health psychology: Developing biologically plausible models linking the social world and physical health." *Annual Review of Psychology*, 60 (2009): 501–524; Teresa E. Seeman, Burton H. Singer, John W. Rowe, Ralph I. Horwitz, and Bruce S. McEwen, "Price of adaptation – allostatic load and its health consequences: MacArthur studies of successful aging." *Archives of Internal Medicine*, 157, no. 19 (1997): 2259–2268.

29 Debra J. Crews and Daniel M. Landers, "A meta-analytic review of aerobic fitness and reactivity to psychosocial stressors." *Medicine & Science in Sports & Exercise*, 19 (1987): S114–20; Anastasia Georgiades, Andrew Sherwood, Elizabeth C.D. Gullette, Michael A., Babyak, Alan Hinderliter, Robert Waugh, Damon Tweedy, Linda Craighead, Richard Bloomer, and James A. Blumenthal, "Effects of exercise and weight loss on mental stress–induced cardiovascular responses in individuals with high blood pressure." *Hypertension*, 36, no. 2 (2000): 171–176.

30 Marcus K. Taylor, Amanda E. Markham, Jared P. Reis, Genieleah A. Padilla, Eric G. Potterat, Sean P.A .Drummond, and Lilianne R. Mujica-Parodi, "Physical fitness influences stress reactions to extreme military training."

Military Medicine, 173, no. 8 (2008): 738–742, https://doi.org/10.7205/MIL MED.173.8.738
31 American Psychological Association. Stress in America. April 30, 2012, http://www.stressinamerica.orgAccessed2013; Rafer S. Lutz, Matthew A. Stults-Kolehmainen, and John B. Bartholomew, "Exercise caution when stressed: stages of change and the stress–exercise participation relationship." *Psychology of Sport and Exercise*, 11, no. 6 (2010): 560–567.
32 Marine Azevedo Da Silva, Archana Singh-Manoux, Eric J. Brunner, Sara Kaffashian, Martin J. Shipley, Mika Kivimäki and Nabi Hermann, "Bidirectional association between physical activity and symptoms of anxiety and depression: The Whitehall II study." *European Journal of Epidemiology*, 27, no. 7 (2012): 537–546.
33 Annika Rosengren, Steven Hawken, Stephanie Ôunpuu, Karen Sliwa, Mohammad Zubaid, Wael A. Almahmeed, Kathleen Ngu Blackett et al., "Association of psychosocial risk factors with risk of acute myocardial infarction in 11,119 cases and 13,648 controls from 52 countries (the INTERHEART study): Case-control study." *The Lancet*, 364, no. 9438 (2004): 953–962; Alan Rozanski, James A Blumenthal, and Jay Kaplan, "Impact of psychological factors on the pathogenesis of cardiovascular disease and implications for therapy." *Circulation*, 99, no. 16 (1999): 2192–2217.
34 Suzanne C. Segerstrom and Gregory E. Miller, "Psychological stress and the human immune system: A meta-analytic study of 30 years of inquiry." *Psychological Bulletin*, 130, no. 4 (2004): 601–630.
35 Sheldon Cohen, David A.J. Tyrrell, and Andrew P. Smith, "Psychological stress and susceptibility to the common cold." *New England Journal of Medicine*, 325, no. 9 (1991): 606–612.
36 Robert M. Sapolsky, "Glucocorticoids, stress, and their adverse neurological effects: Relevance to aging." *Experimental Gerontology*, 34, no. 6 (1999): 721–732; Catherine S. Woolley, Elizabeth Gould, and Bruce S. McEwen, "Exposure to excess glucocorticoids alters dendritic morphology of adult hippocampal pyramidal neurons." *Brain Research*, 531, no. 1–2 (1990): 225–231.
37 Peter D. Harms, Dina V. Krasikova, Adam J. Vanhove, Mitchel N. Herian, and Paul B. Lester, "Stress and emotional well-being in military organizations." In *The role of emotion and emotion regulation in job stress and well being* (2013). Emerald Group Publishing Limited.
38 Jennifer Kavanagh, *"Stress and performance a review of the literature and its applicability to the military."* (2005).
39 Nicola T. Fear, Margaret Jones, Dominic Murphy, Lisa Hull, Amy C. Iversen, Bolaji Coker, Louise Machell et al., "What are the consequences of deployment to Iraq and Afghanistan on the mental health of the UK armed forces? A cohort study." *The Lancet*, 375, no. 9728 (2010): 1783–1797.
40 Harms et al., *"Stress in military organizations,"* (2013).

41 Harris R. Lieberman, Gaston P. Bathalon, Christina M. Falco, Charles A. Morgan, Philip J. Niro, and William J. Tharion, "The fog of war: decrements in cognitive performance and mood associated with combat-like stress." *Aviation, Space, and Environmental Medicine*, 76, no. 7 (2005): C7–C14.
42 Harms et al., *"Stress in military organizations,"* (2013).
43 Ibid.
44 Ibid.
45 Ibid.
46 Saunders et al., *"The effect of stress inoculation training,"* (1996).
47 Lieberman et al., "The fog of war," (2005); Carmen Sandi, "Stress, cognitive impairment and cell adhesion molecules." *Nature Reviews Neuroscience*, 5, no. 12 (2004): 917–930.
48 Arthur W. Combs and Charles Taylor, "The effect of the perception of mild degrees of threat on performance." *The Journal of Abnormal and Social Psychology*, 47, no. 2S (1952): 420–424; Christopher D. Wickens, *Processing resources and attention* (2020). CRC Press; Christopher D. Wickens, "Multiple resources and performance prediction." *Theoretical Issues in Ergonomics Science*, 3, no. 2 (2002): 159–177; Christopher D. Wickens, Justin G. Hollands, Simon Banbury, and Raja Parasuraman. In *Engineering psychology and human performance* (2015). Psychology Press.
49 James A. Easterbrook, "The effect of emotion on cue utilization and the organization of behavior." *Psychological Review*, 66, no. 3 (1959): 183–201.
50 Phillip A. Hancock and J. L. Weaver, "On time distortion under stress." *Theoretical Issues in Ergonomics Science*, 6, no. 2 (2005): 193–211.
51 Dave Grossman and Loren W. Christensen, "On combat: The psychology and physiology of deadly conflict in war and peace." *PPCT Research P* (2004).
52 Alan D. Baddeley, "Working memory." *Science*, 255, no. 5044 (1992): 556–559.
53 Alan D. Baddeley and Graham Hitch, "Working memory." In *Psychology of learning and motivation*, Vol. 8, pp. 47–89 (1974). Academic Press.
54 Pierre Barrouillet, Sophie Portrat, and Valérie Camos, "On the law relating processing to storage in working memory." *Psychological Review*, 118, no. 2 (2011): 175–192.
55 Lieberman et al., *"The fog of war,"* (2005).
56 Ibid.
57 Staal, *Stress* (2004).
58 Giora Keinan, "Decision making under stress: Scanning of alternatives under controllable and uncontrollable threats." *Journal of Personality and Social Psychology*, 52, no. 3 (1987): 639–644.
59 Janet G. Baradell and Kitty Klein, "Relationship of life stress and body consciousness to hypervigilant decision making." *Journal of Personality and Social Psychology*, 64, no. 2 (1993): 267–273.

60 Richard S. Lazarus and Charles W. Eriksen, "Effects of failure stress upon skilled performance." *Journal of Experimental Psychology* 43, no. 2 (1952): 100–105.
61 Hans Selye, "Stress without distress. Philadelphia." (1974): 75–89.
62 Roman Kupriyanov and Renad Zhdanov, "The eustress concept: Problems and outlooks." *World Journal of Medical Sciences*, 11, no. 2 (2014): 179–185; Selye, "Stress," (1974).
63 Lazarus, "Psychological stress," (1993).
64 Le Fevre, Mark, Gregory S. Kolt, and Jonathan Matheny, "Eustress, distress and their interpretation in primary and secondary occupational stress management interventions: Which way first?." *Journal of Managerial Psychology*, (2006): 547–565.
65 Richard S. Lazarus and Elizabeth Alfert, "Short-circuiting of threat by experimentally altering cognitive appraisal." *The Journal of Abnormal and Social Psychology*, 69, no. 2 (1964): 195–205; Joseph C. Speisman, Richard S. Lazarus, Arnold Mordkoff, and Les Davison, "Experimental reduction of stress based on ego-defense theory." *The Journal of Abnormal and Social Psychology*, 68, no. 4 (1964): 367–380.
66 Carlyle H. Folkins, "Temporal factors and the cognitive mediators of stress reaction." *Journal of Personality and Social Psychology*, 14, no. 2 (1970): 173–184; Alan Monat, James R. Averill, and Richard S. Lazarus, "Anticipatory stress and coping reactions under various conditions of uncertainty." *Journal of Personality and Social Psychology*, 24, no. 2 (1972): 237–253.
67 Richard S. Lazarus and Susan Folkman, *Stress, appraisal, and coping* (1984). Springer Publishing Company.
68 Lazarus, *"Psychological stress,"* (1993).
69 Lazarus, *Stress* (1984); Sarah Mae Sincero, *"Stress and cognitive appraisal."* (2012): https://explorable.com/stress-and-cognitive-appraisal.
70 Sincero, "Stress," (2012).
71 Lazarus and Folkman, *Stress* (1984).
72 Sincero, "Stress," (2012).
73 Lazarus and Folkman, *Stress* (1984).
74 David Scott Yeager and Carol S. Dweck, "Mindsets that promote resilience: When students believe that personal characteristics can be developed." *Educational Psychologist* 47, no. 4 (2012): 302.
75 Carol S. Dweck, *Mindset: The new psychology of success* (2008). Random House Digital, Inc.
76 Dweck, *Mindset* (2008).
77 Yeager et al., *"Mindsets that promote resilience,"* (2012).
78 Alia J. Crum and Ellen J. Langer, "Mind-set matters: Exercise and the placebo effect." *Psychological Science*, 18, no. 2 (2007): 165–171; Carol S. Dweck, "Mindsets and math/science achievement." (2014); Becca R. Levy, Jeffrey

M. Hausdorff, Rebecca Hencke, and Jeanne Y. Wei, "Reducing cardiovascular stress with positive self-stereotypes of aging." *The Journals of Gerontology Series B: Psychological Sciences and Social Sciences*, 55, no. 4 (2000): P205–P213; Becca R. Levy, Martin D. Slade, Suzanne R. Kunkel, and Stanislav V. Kasl, "Longevity increased by positive self-perceptions of aging." *Journal of Personality and Social Psychology*, 83, no. 2 (2002): 261–270; Becca R. Levy and Lindsey M. Myers, "Preventive health behaviors influenced by self-perceptions of aging." *Preventive Medicine*, 39, no. 3 (2004): 625–629.

79 Larry Cahill, Lukasz Gorski, and Kathryn Le, "Enhanced human memory consolidation with post-learning stress: Interaction with the degree of arousal at encoding." *Learning & Memory*, 10, no. 4 (2003): 270–274.

80 Elissa S. Epel, Bruce S. McEwen, and Jeannette R. Ickovics, "Embodying psychological thriving: Physical thriving in response to stress." *Journal of Social Issues*, 54, no. 2 (1998): 301–322.

81 Richard G. Tedeschi and Lawrence G. Calhoun, "Posttraumatic growth: Conceptual foundations and empirical evidence". *Psychological Inquiry*, 15, no. 1 (2004): 1–18.

82 Alia Crum, "Evaluating a mindset training program to unleash the enhancing nature of stress." In *Academy of management proceedings*, Vol. 2011, no. 1, pp. 1–6 (2011). Briarcliff Manor, NY 10510: Academy of Management.

83 Alia J. Crum, Modupe Akinola, Ashley Martin, Sean Fath, and Alia J. Crum, "The benefits of a stress-is-enhancing mindset in both challenging and threatening contexts." *Anxiety, Stress and Coping* (2015): 1–25; Crum, Alia J., Modupe Akinola, Ashley Martin, and Sean Fath, "The role of stress mindset in shaping cognitive, emotional, and physiological responses to challenging and threatening stress." *Anxiety, Stress, & Coping*, 30, no. 4 (2017): 379–395.

84 Christopher J. Kilby and Kerry A. Sherman, "Delineating the relationship between stress mindset and primary appraisals: Preliminary findings." *Springerplus*, 5, no. 1 (2016): 1–8.

85 Anne Casper, Sabine Sonnentag, and Stephanie Tremmel, "Mindset matters: The role of employees' stress mindset for day-specific reactions to workload anticipation." *European Journal of Work and Organizational Psychology*, 26, no. 6 (2017): 798–810.

86 Jeremy P. Jamieson, Alia J. Crum, J. Parker Goyer, Marisa E. Marotta, and Modupe Akinola, "Optimizing stress responses with reappraisal and mindset interventions: an integrated model." *Anxiety, Stress, & Coping*, 31, no. 3 (2018): 245–261.

87 Crum, *"Evaluating a mindset training program,"* (2011).

88 Alia J. Crum, Peter Salovey, and Shawn Achor, "Rethinking stress: The role of mindsets in determining the stress response." *Journal of Personality and Social Psychology*, 104, no. 4 (2013): 716.

89 Susan. J. Torres and Caryl A. Nowson, "Relationship between stress, eating behavior, and obesity." *Nutrition*, 23, (2007): 887–894.
90 M.L. Meewisse, Johannes B. Reitsma, Giel-Jan de Vries, Berthold P. R. Gersons, and Miranda Olff, "Cortisol and post-traumatic stress disorder in adults: Systematic review and meta-analysis." *The British Journal of Psychiatry*, 191, (2007): 387–392.
91 Crum, *"The benefits of a stress-is-enhancing mindset."* (2015).
92 Eric N. Smith, Michael D. Young, and Alia J. Crum, "Stress, mindsets, and success in Navy SEALs special warfare training." *Frontiers in Psychology*, 10 (2020): 2962.
93 Kyla Haimovitz and Carol S. Dweck, "Parents' views of failure predict children's fixed and growth intelligence mind-sets." *Psychological Science*, 27, no. 6 (2016): 859–869.
94 Veronika Job, Gregory M. Walton, Katharina Bernecker, and Carol S. Dweck, "Implicit theories about willpower predict self-regulation and grades in everyday life." *Journal of Personality and Social Psychology*, 108, no. 4 (2015): 637.
95 Smith, *"Stress,"* (2020).
96 Crum, *"Rethinking stress,"* (2013).

Edith Knight

Meaningful Work: What Does It Mean?

The Search for Meaning

Across the ages people have contemplated the meaning of life – many have sought spiritual or religious guidance from ancient scriptures, such as the Adi Granth, Bhagavad Gītā, Bible and Qur'an. Scholars have turned to astronomy, philosophy, psychology, and religion to seek a greater understanding of our existence. The ancient Greek philosopher Aristotle wrote about living virtuously to achieve fulfilment and, later in the mid-20th century, psychologist Abraham Maslow wrote about the necessary precursors to achieve self-actualization. These well-known scholars and many before and after them have contemplated how to live a meaningful life. Given that for most people work takes up the largest portion of waking hours, it is no wonder that people also search for meaning in their work. Indeed, Victor Frankl considered meaning to be of primary importance in life and believed that work was a major contributor to it.[1]

Judeo-Christian scholars, and perhaps those of other faiths, consider work as a means to fulfill their spiritual aspirations: a *vocation* or a *calling*, but vocations are increasingly applied in the secular domain to refer to any employment or occupation that is worthy of great dedication. More recently, as health and wellness and positive psychology have taken root in mainstream psychology, researchers have become interested in studying meaningful work more broadly and are striving to understand what meaning signifies to individuals in the workplace.[2]

Different domains may consider *callings* and *meaningful work* as distinct but related constructs. Notwithstanding their distinctions, research suggests that when individuals have a calling or find activities meaningful, they also tend to report more life satisfaction and higher levels of well-being. There are many positive benefits for organizations when workers find their jobs meaningful; employees are more engaged in their work, experience greater job satisfaction and their overall well-being is enhanced.[3] Considering these benefits, it seems pertinent to ask: How can organizations infuse meaning into the workplace and what practical interventions might contribute to more meaningful work?

In this chapter a brief overview of the historical and theoretical background of two constructs, calling and meaningful work, is provided along with definitions, antecedents, outcomes, and arguments for why military leaders should care about these concepts. Some activities the Canadian Armed Forces is engaged in that foster meaningful work will be presented, followed by some practical advice intended to guide organizations in the development of initiatives that will encourage employees to find meaning, such as: transformational leadership, interfaith capacity building, job design, vocational counselling, tailored supervisory feedback, and job crafting. These techniques have the potential to enhance meaning, spirituality, creativity, productivity, and most importantly, well-being in the workplace.

What Does It Mean to Be Called?

The concept of calling has evolved a great deal over the years. Early European origins of the concept date back to the 16th century when Protestants, Martin Luther and John Calvin, advanced the idea that occupations can have spiritual significance.[4] A recent comprehensive definition of calling is,

> "...a transcendent summons, experienced as originating beyond the self, to approach a particular life role in a manner oriented toward demonstrating or deriving a sense of purpose or meaningfulness and that holds other-oriented values and goals as primary sources of motivation."[5]

This definition includes three components: a transcendent source, purpose or meaning, and pro-social motivation. Transcendent summons implies that calling is derived from a source external to the individual, though other definitions propose that one's calling may be inspired from internal sources aligned with personal moral convictions that may not necessarily be faith-based, such as the one below:

> "...a course of action in pursuit of pro-social intentions embodying the convergence of an individual's sense of what he or she would like to do, should do, and actually does."[6]

The study of callings in the Occident has been extended to secular populations to capture those who may seek fulfilment or meaning from non-religious sources in life and in work. Some of these scholars argued that earlier definitions of calling were too contextual, lacking empirical support and adequate measures and proposed the following definition:

> "...a consuming, meaningful passion people experience toward a domain."[7]

The aforementioned definitions are broad enough to encapsulate secular views but lack an occupation-related focus. Two of the definitions include the concept of meaning derived either from a '*life role*' or a '*domain*', suggesting that meaning may be an important part of having a calling, but not necessarily within an occupation. Furthermore, it remains to be specified if the meaning contained within the calling construct is global or contextual (i.e., meaning in *life*, meaning in *work*, or *both*). The fact that many definitions of calling include the word *meaning* has also made it difficult for researchers to disentangle the concepts of calling and meaningful work.[8]

There have been recent attempts to bring more conceptual clarity to the field of research on callings, especially given its proliferation over the last 20 years.[9] Despite the limitations and lack of consensus in refining the concept, researchers in the Occident who endorse various definitions agree that having a calling helps guide people toward satisfying life and work.[10] Callings are associated with work that is socially valuable and personally rewarding, though not always pleasant or profitable and are considered an evolving process which is discovered over time as attributions are made about the meaning and purpose of activities.[11] Callings tend to have a religious or spiritual origin, but have also been defined more simply as roles that are useful to society and bring a sense of fulfilment. In both cases, callings have been found to be distinct from having a career, a job, a vocation, and from meaning in life. Given the wide range of definitions, it is not surprising that those who experience callings are prevalent in society and come from a wide spectrum of occupations from administrative assistants to medical doctors.[12]

Similar to the body of literature on callings, the concept of meaningful work has taken different approaches with nuances in the way terms are used, for example: *meaningful* work, *meaningfulness* of work, meaning *of* work and meaning *in* work are subtly different in their conceptualizations and will be explored next.[13]

Meaningful Work

Around the mid-1900s, psychologists Carl Jung and Sigmund Freud, were of the opinion that *meaningful work* was important for psychological wellness and to realize one's full potential.[14] In 1975, and prior to more recent explorations of vocations and callings in the workplace, seminal work was developed on job characteristics and job design by Hackman and Oldham. Meaningful work was purported to be one of three critical elements to foster work productivity, along with providing employees responsibility for outcomes and knowledge of the results of their work tasks. They found that when these conditions were met, the results were positive motivation toward work, higher job satisfaction, higher levels of performance and lower levels of absenteeism and turnover.[15]

A multidimensional measure of *meaningful work* (MW) narrowed the concept further:[16]

> "...we define MW not as simply whatever work means to people (meaning), but as work that is both significant and positive in valence (meaningfulness). Furthermore, we add that the positive valence of MW has a eudaimonic (growth- and purpose-oriented) rather than hedonic (pleasure-oriented) focus."

The second approach, *meaningfulness of work* has been defined in terms of evaluations employees make about their work experience and the degree of congruence that exists between their valuation of work and their self-concept:

> "...meaningfulness, defined as the perceived value of one's job in relation to one's personal beliefs, attitudes, and values."[17]

The terms *meaningfulness of work* and *meaning of work* are related concepts that have been used interchangeably in the field of psychology, but may differ, as pointed out by one researcher who considers the field of sociology, where the source of *meaning of work* can evolve from a social context, not just from the individual.[18] *Meaning of work* can refer to a type of meaning which can have a positive or negative connotation (e.g., to socialize or as a necessity to survive), whereas meaningfulness of work is defined in a positive way.

Other researchers differentiated *meaning of work* from *meaning in work*, such that *meaning in work* is a subjective personal phenomenon that occurs in various work contexts with an experiential focus, and *meaning of work* can be examined more globally (i.e., work as a social institution), but on different levels (i.e., personal, organizational, or societal).[19] The concept of *meaning in work* comprises four factors: coherence (i.e., degree of fit between self-concept and role at work), direction (i.e., worker/corporate value alignment), significance (i.e., opportunities for societal contribution), and belonging (i.e., positive social work environment).

The concept of *meaning in work* is the most comprehensive, but still lacks clarity as it shares some overlap with theoretical foundations of meaningful work and uses the term meaningful work as the outcome measure in its model.[20] Selecting optimal definitions of meaning and calling is a first step to disentangling these two similar concepts. A second step requires empirically testing these definitions in an applied work setting. As acknowledged in the previous study, meaning is often included within definitions of calling, so future research should attempt to distinguish these related constructs to provide insights into whether meaning (in work) and calling are different constructs and if calling might be a precursor to meaningful work.

As with the calling literature, large-scale reviews have attempted to consolidate the knowledge base on meaningful work to guide future research and identify where the gaps are. An analysis of empirical research identified several different perspectives on meaningful work including: psychological (including eudaimonic), spiritual, humanistic, and occupational; among the most widely used theories identified were Hackman and Oldham's job characteristics model (previously mentioned) and transformational leadership theory (i.e., wherein leaders can enhance meaningfulness of work).[21] The authors concluded that the most comprehensive models of meaningful work included both self-and other-oriented experiences (i.e., self-development and belongingness), but that there has not yet been sufficient development of alternative approaches and methodologies within the empirical literature to draw conclusions on.

Joining the Concepts – Meaningful Work and Calling

Though both fields of research lack conceptual clarity, each of their most comprehensive definitions share at least two elements when applied to the workplace: (1) both are an employee's subjective positive evaluation that drives their motivation; and (2) both include some element of societal contribution. Various employee standpoints, whether they be agnostic, spiritual, or religious, will influence perceptions of what work means and whether or not it is considered a calling, meaningful work, or something else. Regardless of perspective, when work is aligned with personal values, is part of one's identity, permits feelings of belongingness, contributes positively to society, is considered significant in one's life, and contributes to self-esteem, the outcomes are bound to be positive for both the organization and the individual, no matter the definition applied. As organizations around the world become increasingly diverse, it will be important to identify organizational constructs that resonate across varied perspectives.

Precursors to Meaningful Work and Calling

How does work become meaningful? How do callings develop? Are there individual characteristics, social or environmental work conditions that promote the development of meaningful work and/or callings? To answer these questions, we will first look at personal characteristics commonly associated with those who perceive themselves as having meaningful work and those who develop callings.

Research in the Occident suggests that individuals need to be actively searching for meaning before they can develop a calling.[22] Those with a calling tend to be curious in nature, have a clear sense of self and future career aspirations, with vocational identity, and have a sense of purpose or meaning in life.[23] Those with callings have also been associated with greater psychological capital. For instance, in a time-lagged cross sectional survey of university students and faculty, those with characteristics of self-efficacy, hope, optimism and resilience (i.e., psychological capital) were more likely to respond positively to having answered their calling.[24] In addition, those with proactive personalities were more likely to experience occupational callings with curiosity playing a role in strengthening those relationships.[25] Proactive personality included people who are self-starters, future-focused, curious and persistent.

An examination of over 500 teachers in South Africa looked at several important variables to determine the precursors and outcomes of meaningful work.[26] Meaningful work was assessed together with work orientation that determined whether teachers considered their work to be a job, a career, or a calling. Job design was assessed through perceptions of the following: autonomy (independence to do the job in a desired way), task identity (being able to complete a task from start to finish), skill variety (level of complexity), task significance (relevance beyond the task itself) and feedback (awareness of how well the job is performed). Other variables assessed were: co-worker relationships, engagement, self-ratings of performance, personal resources, burnout and turnover. Results demonstrated that having a calling orientation, good job design and positive co-worker relations were associated with high ratings of meaningful work, which in turn predicted work engagement. The authors explain that teachers with calling orientations may be better in touch with their identities and feel a strong sense of purpose in the performance of their jobs and when combined with positive co-worker relationships and good job design, the workplace environment supports a better work experience.

Similarly, another study examined the precursors to meaningful work and identified three important variables: work-role fit, task significance and socio-moral climate. Specifically, work-role fit is the degree to which people identify with the work they perform (analogous to having a calling orientation); significance of work tasks refers to the impact the task can have for the organization or for society as a whole; and socio-moral climate refers to the way employees are able to collaborate and communicate in a respectful and ethical environment – all variables contributed to perceptions of meaningful work.[27]

Other researchers have examined three precursors to meaningful work: fairness, responsible leadership, and worthy work.[28] Fairness was measured with a scale of distributive justice that taps into pay equity. Responsible leadership assessed the degree to which leaders displayed inspirational behaviours in the workplace related to authenticity and ethical conduct. Worthy work referred to jobs that were pro-social or pro-environmental (e.g., improves the environment, enriches people's lives by promoting wellness, alleviates suffering, eliminates danger, pollutions). Results demonstrated that worthy work was the most dominant predictor of meaningful work of the three variables, but that all three did predict some aspects of meaningful work.

Outcomes of Meaningful Work and Callings

Positive Outcomes

Research shows that meaning is important to individuals' well-being whether it originates from a career or from life in general.[29] Meaning then, casts a broad spectrum that includes both conceptualizations of calling and meaningful work. Indeed, when employees are searching for meaning in life and find meaning through work, they tend to experience improved well-being and more confidence in their career decisions.[30]

A recent meta-analysis spanning 20 years of research on calling in the Occident found there were positive benefits for individuals and organizations such that those who perceived themselves as having a calling experienced higher levels of job and domain satisfaction and career commitment than those without a calling.[31] Those who felt called were also more committed to their careers than those without a calling. Furthermore, those individuals with callings experienced both a higher degree of psychological and subjective well-being than those without.

Correspondingly, a review of the meaning *in* work literature shows that many people are searching for a greater sense of meaning in their work (e.g., employees in education, health care, finance, science and technology, and retail); and when they find it, it is associated with higher levels of work engagement, job satisfaction, performance, organizational commitment, and tenure.[32] Furthermore, meaningful work has been found to enrich peoples' lives by creating a spillover effect, such that being energized at work and having a positive mood influence experiences outside of the workplace (e.g., home life, personal activities).[33]

Adverse Outcomes

The study with South African teachers (previously mentioned) found that those with low calling orientations and poor co-worker experiences were more likely to experience burnout, and those lacking a calling orientation were more likely to turnover.[34] Similarly, other research demonstrates that having a calling results in fewer days missed at work, fewer withdrawal intentions and fewer depressive symptoms.[35] These correlational studies preclude us from understanding cause-effect relationships (e.g., does low calling result in burnout or does burnout result in low calling?), but the results do mirror research on meaning in work that is associated with lower levels of turnover intentions, disengagement, cynicism, exhaustion, and stress.[36]

To address the limitations of correlational research, a longitudinal study with aspiring professional musicians and medical students suggests that perceptions of calling fluctuate over time as a result of environmental stressors.[37] Life experiences influence values and beliefs as individuals develop and evolve, so it is reasonable to expect that one's feelings about work and career will change over time. The ideals of youth are likely to erode when life's experiences challenge preconceived assumptions. This has implications for the workplace, because both stressors and unrealistic expectations may reduce levels of calling/meaning which may ultimately reduce levels of employee commitment, satisfaction, and desire to remain with the organization. Other research has demonstrated that when career expectations are unmet, employees are more likely to turnover.[38]

Why Military Leaders Should Care

1. Meaningful Work May Help Buffer the Effects of Post-Deployment Illness

Archival data from a 2010 survey of deployed U.S. Air Force medical personnel was examined and revealed that those who had experienced combat exposure were more likely to report sleep problems and interpersonal withdrawal; those who reported difficulty during their deployment experienced more depression and problems at work (i.e., decreased professional skills), as well as sleep problems and interpersonal withdrawal.[39] Interestingly, those who scored high in meaningful work experienced fewer work-related health issues after their deployment.

2. Finding Positive Meaning from Adverse Events Is Associated with Improved Mental Health

The role of meaning was examined in a sample of 3,000 male Dutch veterans who had participated in war and peacekeeping missions.[40] This study supports the notion that when exposure to war-zone stressors is linked with perceptions of high threats, one's global feelings of trust in others and in the world are negatively impacted and associated with more intrusive thoughts that are linked to reduced mental health and lower quality of life. On the other hand, those who seek to re-establish trust in others and find positive meaning from their traumatic experience are more satisfied with their mental health and quality of life. As explained by researchers in this study, after a traumatic experience people look for meaning by reframing their perspectives and attitudes and by seeking explanations for why the event occurred as part of a psychological process. Indeed, the term post-traumatic growth (PTG) was coined for those who thrive after experiences of trauma.[41] PTG is an individualized process that occurs to those who are able to realize a greater appreciation for life by realigning priorities and seeking meaning from life's adverse events.[42]

How the Canadian Armed Forces (CAF) Promotes Meaningful Work

Canada's Defence Policy: *Strong, Secure, Engaged* (SSE), launched in 2017, announced a comprehensive wellness strategy that includes gender, diversity and inclusion initiatives that enable every member of the Defence Team to make meaningful contributions. Following that, the *Total Health and Wellness Strategic Framework* was published, which included a spiritual dimension at the core of the wellness concept that recognizes individuals' needs for physical, mental, and spiritual health not only at home but in the workplace.[43] There are benefits for organizations when they welcome diversity and encourage interfaith discourse. For example, when people are free to discuss their religious perspectives in the workplace, they tend to experience higher levels of job satisfaction and well-being.[44]

Aligned with SSE and the *Total Health and Wellness Strategy*, the Royal Canadian Chaplain Service (RCChS) supports the spiritual well-being of all military personnel and their families regardless of faith, belief or custom.[45] The RCChS launched a spiritual wellness strategy in 2017 that describes spiritual wellness as that which gives meaning and purpose to life and is central to the development of moral character, values and beliefs, and is intrinsic to how individuals experience self, others, and community.[46] At the time of this writing, the RCChS had successfully recruited 9 Muslim Chaplains and has enrolled Chaplains of the Buddhist, Christian, Jewish, Sikh, and Unitarian Universalism faith traditions to provide improved interfaith services.

To assess psychological health and safety among approximately 40,000 Defence Team members (comprised of Regular Force, Reserve Force, and civilians), Director General Military Personnel Research and Analysis administered the *Defence Workplace Well-being Survey* in 2018. Results from 13,112 respondents demonstrated that meaningful work, perceived organizational support, relatedness (social support from colleagues) and job stress were all significant drivers of workplace well-being. More specifically, high levels of job stress combined with low levels of organizational and social support and a lack of meaning in work were associated with lower workplace well-being.[47]

Practical Interventions

How Can Leaders Make Work More Meaningful?

Leaders Can Work with Their Followers to Design More Personally Meaningful Jobs

The way a job is designed impacts a worker's experience and contributes to perceptions of meaningful work.[48] Providing a variety of tasks makes work more challenging and interesting and allows employees to use more of their skills and talents.[49] Also, when employees are able to see a project through from beginning to end and are able to realize a tangible outcome, they are provided with a sense of ownership and satisfaction through what is referred to as task identity.[50] When outcomes of work are more abstract or shared with other parts of an organization, thwarting task identity, it may help employees find task significance when leaders can demonstrate the relevance of their work by explaining process flow and how followers' contributions result in a final outcome – in this way leaders help employees realize the value of their work.[51]

2. Leaders Can Enhance Meaning in Work by Fostering Psychological Work Needs

There are three psychological work needs required for optimal human development, as explained by Self-Determination Theory (SDT), that underlie intrinsic motivation: competence, autonomy, and relatedness.[52] First, leaders contribute to follower's self-perceptions of competence by providing opportunities for training and professional development. Leaders also enhance meaning in work by providing feedback on how well a job is done and pointing out the task's significance for the attainment of organizational goals. Second, when leaders provide their employees autonomy to do their jobs in a way that permits application of skills, creativity and use of professional judgment, it also provides more opportunities for employees to enhance meaning in their own work, such as through job crafting.[53] Job crafting is a concept first put forward in 2001 to describe the process employees use to initiate changes to their jobs to take advantage of their personal knowledge, skills, talents and interests to enhance person-job fit.[54] Job crafting has been shown to improve engagement at work, productivity, and also the well-being of coworkers.[55]

The third important psychological work need is relatedness; leaders can enhance interpersonal working relationships among co-workers by facilitating collaborative working arrangements, encouraging development of communication skills, and by organizing team social events that celebrate diversity and recognize both individual and team accomplishments. Leaders who embrace cultural diversity facilitate a workplace environment where employees feel free to celebrate and share their unique ethnicities and faith affiliations. Cultural identity is thought to be developed and maintained through participation in culturally meaningful activities that offer opportunities to form interracial social connections which then contribute to feelings of belongingness. When marginalized groups have opportunities to participate in personally meaningful culture-related activities, they experience feelings of significance and purpose which help counteract negative psychological symptoms (depression) and reduces feelings of worthlessness, social alienation, loneliness, and isolation.[56]

3. Leaders Can Use Transformational Leadership Behaviours

Transformational leaders inspire their followers and promote a sense of purpose that contributes to enhanced meaning in work, contributing to higher levels of employee engagement.[57] For example, results from a study involving managers and top executives across different industries revealed that managers with positive follower characteristics (i.e., creative, innovative, proactive, showing initiative, and having a learning orientation) were more engaged in their work when their leaders (top executives) displayed transformational leadership qualities.[58] Transformational leadership is widely accepted as a concept and is linked with well-being, task performance, and creativity; these leaders are visionary mentors who motivate and develop followers through their ability to influence and draw out creativity and innovative ideas.[59]

Leaders play an important role in enabling employee involvement through their interpersonal interactions, and by doing so can bring forward novel ideas and valuable contributions from their followers.[60] By practicing active listening, leaders provide subordinates opportunities to share their ideas, which then contributes to feelings of being valued. When followers do not feel heard or if they feel unable to make valuable contributions through their work, they may become cynical, feel unfulfilled and yearn for more meaningful engagement. Alternatively, when employees have positive interpersonal engagements with their leaders, research shows increases in extra-role behaviors, reduced psychological strain and fewer turnover intentions.[61]

How Can Organizations Make Work More Meaningful?

Person-Job Fit

Research has shown that person-job fit is important for work to be meaningful.[62] When employees' interests and skills are aligned with work (coherence), they experience higher levels of meaningful work as well as professional efficacy (i.e., feelings of competence and satisfaction with work achievements).[63] Interest inventories aid job seekers to understand their own skills, strengths and interests and also help organizations with job placement by identifying career fields aligned with individuals' interests. Use of interest inventories facilitates improved recruiting and selection processes, thereby optimizing person-job fit. Optimal placement of employees at the onset of onboarding is cost effective, as it can reduce internal movement and re-training costs and contributes to meaningful work as well as job satisfaction, which is associated with increased commitment and retention.[64]

Psychologically Healthy Workplaces

Researchers and practitioners have recognized the importance of psychologically healthy workplaces that take into account a multitude of factors important for well-being.[65] For example, physical work environment (infrastructure, building design, air quality, lighting), psychosocial work environment (leadership, social culture, meaningful work, job design), and individual factors (personality, mental, spiritual and physical health) are all contributors to psychologically healthy workplaces.[66] Development of broad strategic interventions that treat healthy workplaces in a holistic manner are shown to be more effective than those with a narrow scope.[67]

Interfaith Capacity Building

Guidelines and best practices for interfaith capacity building are available to help organizations facilitate healthy interpersonal relationships among diverse groups.[68] One successful interfaith capacity building project brought together Muslim and Jewish youth groups for sustained engagement in a large city in the U.K. Three critical components included: (1) creating a space where participants from different groups could safely come together to socialize; (2) establishing shared interests (cricket); and (3) creating informal time to develop a shared identity (hanging out).[69] Embracing diversity means more than filling employment equity targets. Organizations must adopt a worldview perspective and implement initiatives that positively influence workplace culture by celebrating diversity and removing discriminatory practices. By fostering relatedness to others in the workplace, organizations fulfill important psychological work needs and provide a safe place to express identity that facilitates creativity and contributes to meaningful work.[70]

Conclusion and Recommendations to Readers

The literature on callings and meaningful work requires more clarity to advance conceptual development and research in the area. However, both concepts have demonstrated their value and we know that both callings and meaningful work are of primary importance in people's lives and contribute both to meaningful life and work and to overall well-being. Organizations can implement psychologically healthy workplace practices that include celebration of diversity to reap the many benefits associated with healthy and meaningful workplaces: higher levels of job satisfaction, engagement, commitment, and reduced levels of stress, exhaustion, cynicism, and turnover intentions. Organizations can also save re-training costs by using interest inventories to optimize person-job-fit, thereby enhancing job satisfaction and meaning in work.

As workspaces become more diverse, it behooves leaders and organizations to embrace cultural diversity by removing discriminatory practices and creating safe environments where all employees feel free to share and celebrate their unique ethnicities and faith affiliations. Leaders have a unique position in the organization where they directly interact with employees and, therefore, have an immediate and profound impact on their experience in the workplace. Using transformational leadership behaviors, leaders can inspire and motivate their followers to reach their potential and help them find meaning in their work. By providing clear expectations, reasonable timelines and fostering a socially inclusive work environment where collaboration and job crafting are encouraged, leaders inject meaning into the workplace and help their followers thrive by satisfying their psychological work needs, thereby enhancing the overall employee experience and contributing to a healthy productive workplace.

Notes

1 Victor Frankl, *Man's search for meaning* (2006). Boston: Beacon Press.
2 Brent D. Rosso, Kathryn H. Dekas, and Amy Wrzeniewski, "On the meaning of work: A theoretical integration and review," *Research in Organizational Behavior*, 30 (2010): 91–127, https://doi.org/10.1016/j.riob.2010.09.001

3 Ryan D. Duffy, Bryan J. Dik, and Michael F. Steger, "Calling and work-related outcomes: Career commitment as a mediator," *Journal of Vocational Behavior*, 78, no. 2 (2011): 210–218, https://doi.org/10.1016/j.jvb.2010.09.013
4 For historical accounts, see: Bryan J. Dik and Ryan D. Duffy, "Calling and vocation at work," *The Counseling Psychologist*, 37, no. 3 (2009): 424–450, https://doi.org/10.1177/0011000008316430; Richard Treadgold, "Transcendent vocations: Their relationship to stress, depression and clarity of self-concept," *Journal of Humanistic Psychology*, 39, no. 1 (Winter 1999): 81–105.
5 Dik and Duffy, *"Calling and vocation,"* 427.
6 A.R. Elangovan, Craig C. Pinder, and Murdith McLean, "Callings and organizational behavior," *Journal of Vocational Behavior*, 76, no. 3 (June 2010): 430, https://doi.org/10.1016/j.jvb.2009.10.009
7 Shoshana R. Dobrow and Jennifer Tosti-Kharas, "Calling: The development of a scale measure," *Personnel Psychology*, 67, no. 4 (November 2011): 1003, https://doi.org/10.1111/j.1744-6570.2011.01234.x
8 Dobrow and Tosti-Kharas, *"Calling: A scale measure,"* 1001–1049.
9 Bryan J. Dik and Adelyn B. Shimizu, "Multiple meanings of calling: Next steps for studying an evolving construct," *Journal of Career Assessment*, 27, no. 2 (2019): 323–336, https://doi.org/10.1177/1069072717748676
10 Elangovan, Pinder, and McLean, *"Callings and organizational behavior,"* 428–440.
11 Amy Wrzesniewski, Clark R. McCauley, Paul Rozin, and Barry Schwartz, "Jobs, careers, and callings: People's relations to their work," *Journal of Research in Personality*, 31, no. 1 (1997): 21–33, https://doi.org/10.1006/jrpe.1997.2162; Dik and Duffy, "Calling and vocation," 424–450.
12 Wrzesniewski, McCauley, Rozin, and Schwartz, *"Jobs, careers, and callings,"* 21–33.
13 Tatjana Schnell, Thomas Höge, and Edith Pollet, "Predicting meaning in work: Theory, data, implications," *The Journal of Positive Psychology*, 8, no. 6 (2013): 543–554, https://doi.org/10.1080/17439760.2013.830763
14 Treadgold, *"Transcendent vocations,"* 81–105.
15 J. Richard Hackman and Greg R. Oldham, "Development of the job diagnostic survey," *Journal of Applied Psychology*, 60, no. 2 (1975): 159–170.
16 Michael F. Steger, Bryan J. Dik, and Ryan D. Duffy, "Measuring meaningful work: The Work and Meaning Inventory (WAMI)," *Journal of Career Assessment*, 20, no. 3 (2012): 323, https://doi.org/10.1177/1069072711436160
17 Marylène Gagné, Caroline B. Senécal, and Richard Koestner, "Proximal job characteristics, feelings of empowerment, and intrinsic motivation: A multidimensional model," *Journal of Applied Social Psychology*, 27, no. 14 (1997): 1223.
18 Brent D. Rosso, Kathryn H. Dekas, and Amy Wrzeniewski, "On the meaning of work: A theoretical integration and review," *Research in Organizational Behavior*, 30 (2010): 91–127, https://doi.org/10.1016/j.riob.2010.09.001

19 Schnell, Höge, and Pollet, *"Predicting meaning in work,"* 543–554.
20 Ibid.
21 Catherine Bailey, Ruth Yeoman, Adrian Madden, Marc Thompson, and Gary Kerridge, "A review of the empirical literature on meaningful work: Progress and research agenda," *Human Resource Development Review*, 18, no. 1 (2019): 83–113, https://doi.org/10.1177/1534484318804653
22 Elangovan, Pinder, and McLean, *"Callings and organizational behavior,"* 428–440.
23 Dik and Duffy, "Calling and vocation," 424–450; Yu Guo, Yanjun Guan, Xuhua Yang, Jingwen Xu, Xiang Zhou, Zhuolin She, Peng Jiang, Yang Wang, Jingzhou Pan, Yufan Deng, Ziyue Pan, and Mengyao Fu, "Career adaptability, calling and the professional competence of social work students in China: A career construction perspective," *Journal of Vocational Behavior*, 85, no. 3 (December 2014): 394–402, https://doi.org/10.1016/j.jvb.2014.09.001; Chunyu Zhang, Andreas Hirschi, Anne Herrmann, Jia Wei, and Jinfu Zhang, "The future work self and calling: The mediational role of life meaning," *Journal of Happiness Studies*, 18 (2017): 977–991, https://doi.org/10.1007/s10902-016-9760-y; Jacob A. Galles and Janet G. Lenz, "Relationships among career thoughts, vocational identity, and calling: Implications for practice," *The Career Development Quarterly*, 61, no. 3 (2013): 240–248, https://doi.org/10.1002/j.2161-0045.2013.00052.x
24 Hina Jaffery and Ghulam Abid, "Occupational calling: Antecedents, consequences and mechanism," *Iranian Journal of Management Studies (IJMS)*, 13, no. 3 (2020): 413–439, https://doi.org/10.22059/ijms.2020.254911.673067
25 Jaffery and Abid, *"Occupational calling,"* 413–439.
26 Elmari Fouché, Sebastiaan (Snr) Rothmann, and Corne van der Vyver, "Antecedents and outcomes of meaningful work among school teachers," *SA Journal of Industrial Psychology*, 43 (2017): 1–10, https://doi.org/10.4102/sajip.v43i0.1398
27 Schnell, Höge, and Pollet, *"Predicting meaning in work,"* 543–554.
28 Marjolein Lips-Wiersma, Jarrod Haar, and Sarah Wright, "The effects of fairness, responsible leadership and worthy work on multiple dimensions of meaningful work," *Journal of Business Ethics*, 161 (2020): 35–52, https://doi.org/10.1007/s10551-018-3967-2
29 Michael F. Steger and Bryan J. Dik, "If one is looking for meaning in life, does it help to find meaning in work?," *Applied Psychology: Health and Well-Being*, 1, no. 3 (2009): 303–320, https://doi.org/10.1111/j.1758-0854.2009.01018.x
30 Steger and Dik, *"If one is looking,"* 303–320.
31 Shoshana Dobrow Riza, Hannah Weisman, Daniel Heller, and Jennifer Tosti-Kharas, "Calling attention to 20 years of research: A comprehensive meta-analysis of calling," *Academy of Management Proceedings*, 1 (August 2019) https://doi.org/10.5465/AMBPP.2019.199

32 Schnell, Höge, and Pollet, *"Predicting meaning in work,"* 543–554.
33 Matthew J. Johnson, and Lixin Jiang, "Reaping the benefits of meaningful work: the mediating versus moderating role of work engagement," *Stress Health*, 33, (2016): 288–297, https://doi.org/10.1002/smi.2710
34 Fouché, Rothmann, and van der Vyver, *"Antecedents and outcomes,"* 1–10.
35 Steger and Dik, *"If one is looking,"* 303–320.
36 Schnell, Höge, and Pollet, *"Predicting meaning in work,"* 543–554.
37 Shoshana R. Dobrow, "Dynamics of calling: A longitudinal study of musicians," *Journal of Organizational Behavior*, 34 (2013): 431–452, https://doi.org/10.1002/job.1808; Ryan D. Duffy, R. Stephen Manuel, Nicole J. Borges, and Elizabeth M. Bott, "Calling, vocational development, and well being: A longitudinal study of medical students," *Journal of Vocational Behavior*, 79, no. 2 (2011): 361–366, https://doi.org/10.1016/j.jvb.2011.03.023
38 Inge Houkes, Peter P.M. Janssen, Jan de Jonge, and Arnold B. Bakker, "Specific determinants of intrinsic work motivation, emotional exhaustion and turnover intention: A MuJ., & Bakker, A.B. (2003). Specific determinants of intrinsic work motivation, emotional exhaustion and turnover intention: A multisample longitudinal study," *Journal of Occupational and Organizational Psychology*, 76, no. 4 (December 2003): 427–450, https://doi.org/10.1348/096317903322591578
39 COL Anderson B. Rowan, LTC Wendy J. Travis, Camerson B. Richardson, and LTC Travis R. Adams, "Military mental health personnel deployment survey: a secondary analysis," *Military Medicine*, 185 (2020): 340–346, https://academic.oup.com/milmed/article-abstract/185/3-4/e340/586483.
40 Michaela L. Schok, Rolf J. Kleber, Gerty J.L.M. Lensvelt-Mulders, Martin Elands, and Jos Weerts, "Suspicious minds at risk? the role of meaning in processing war and peacekeeping experiences," *Journal of Applied Social Psychology*, 41, no. 1 (2011): 61–81.
41 Eranda Jayawickreme and Laura E.R. Blackie, "Post-traumatic growth as positive personality change: Evidence, controversies and future directions," *European Journal of Personality*, 28 (2014): 312–331, https://doi.org/10.1002/per.1963
42 Shelby Rodden-Aubut and Jill Tracey, "An uphill battle: A qualitative case study of growth experiences of veterans in the Canadian Armed Forces," *Military Behavioural Health*, 9, no. 1 (2021): 46–54, https://doi.org/10.1080/21635781.2020.1784324
43 Gareth J.M. Doherty, Edith C. Knight, and Tzvetanka Dobreva-Martinova, *"Defence Team Total Health and Wellness Strategic Framework."* In Kelly Farley, & Stefan Sammito (Chairpersons), HFM-302 Symposium on Evidence-based Leader Interventions for Health and Wellness," (Paper presented at the Symposium Conducted at the Meeting of NATO Science and Technology Organization's Human Factors and Medicine Panel, Berlin, Germany, April

2019), https://www.sto.nato.int/publications/STO%20Meeting%20Proceedings/STO-MP-HFM-302/MP-HFM-302-07.pdf

44 Matthew Etherington, "Religion as a workplace issue: A narrative inquiry of two people-one Muslim and the Other Christian," *SAGE Open* (2019): 1–13, https://doi.org/10.1177/2158244019862729

45 https://www.canada.ca/en/department-national-defence/programs/royal-canadian-chaplain.html

46 Sylvain Maurais, *Called to serve 2.0: A spiritual wellness strategy for the Canadian military community* (2017). Ottawa, Canada: Royal Canadian Chaplain Service; This strategy was formalized and updated in 2020, now entitled, *Called to serve (2020–2030): The royal Canadian chaplain service spiritual resilience and well-being strategy* (2020). Ottawa, Canada.

47 Ann-Renee Blais, Glen Howell, Tzvetanka Dobreva-Martinova, Max Hlywa, et al., (Unpublished), *2018 Defence Workplace Well-Being Survey* L0 SharePoint Site (Unpublished report). Ottawa, ON: Defence Research and Development Canada.

48 Fouché, Rothmann, and van der Vyver, "Antecedents and outcomes," 1–10.

49 Hackman and Oldham, "Job diagnostic survey," 159–170.

50 Ibid.

51 Ibid.

52 Marylène Gagné and Edward L. Deci, "Self-determination theory and work motivation," *Journal of Organizational Behavior*, 26 (2005): 331–362, https://doi.org/10.1002/job.322

53 Gagné and Deci, "Self-determination theory," 331–362.

54 Amy Wrzesniewski and Jane E. Dutton, "Crafting a job: Revisioning employees as active crafters of their work," *Academy of Management Review*, 26, no. 2 (2001): 179–201, https://doi.org/10.5465/AMR.2001.4378011

55 Arnold B. Bakker, Maria Tims, and Daantje Derks, "Proactive personality and job performance: The role of job crafting and work engagement," *Human Relations*, 65, no. 10 (2012): 1359–1378, https://doi.org/10.1177/0018726712453471; Maria Tims, Arnold B. Bakker, and Daantje Derks, "Examining job crafting from an interpersonal perspective: Is employee job crafting related to the well-being of colleagues?," *Applied Psychology: An International Review*, 64, no. 4 (October 2015): 727–753, https://doi.org/10.1111/apps.12043

56 Junhyoung Kim, May Kim, Areum Han, and Seungtae Chin, "The importance of culturally meaningful activity for health benefits among older Korean immigrant living in the United States," *International Journal of Qualitative Studies on Health and Well-Being* 10 (2015): 1–9, https://dx.doi.org/10.3402/qhw.v10.27501

57 Antje Schmitt, Deanne N. Den Hartog, and Frank D. Belschak, "Transformational leadership and proactive work behaviour: A moderated mediation model including work engagement and job strain," *Journal of*

Occupational and Organizational Psychology, 89 (2016): 588–610, https://doi.org/10.1111/joop.12143

58 Weichun Zhu, Bruce J. Avolio, and Fred O. Walumbwa, "Moderating role of follower characteristics with transformational leadership and follower work engagement," *Group & Organization Management*, 34, no. 5 (October 2009): 590–619, https://doi.org/10.1177/1059601108.331242.

59 Tony Bush, "Transformational leadership: Exploring common conceptions," *Educational Management Administration & Leadership*, 46, no. 6 (2018): 1–5, https://doi.org/10.1177/1741143218795731; Karina Nielsen, Raymond Randall, Joanna Yarker, and Sten-Olof Brenner, "The effects of transformational leadership on followers' perceived work characteristics and psychological well-being: A longitudinal study," *Work & Stress*, 22, no. 1 (2008): 16–32, https://doi.org/10.1080/02678370801979430; Tony Bush, "Transformational leadership: Exploring common conceptions," *Educational Management Administration & Leadership*, 46, no. 6 (2018): 1–5, https://doi.org/10.1177/1741143218795731

60 Joel B. Carnevale, Lei Huang, Marcus Crede, Peter Harms, and Mary Uhl-Bien, "Leading to stimulate employees' ideas: A quantitative review of leader-member exchange, employee voice, creativity, and innovative behavior," *Applied Psychology: An International Review*, 66, no. 4 (2017): 517–552, https://doi.org/10.1111/apps.12102

61 Dianhan Zheng, Hao Wu, Robert Eisenberger, Lynn M. Shore, Lois E. Tetrick, and Louis C. Buffardi, "Newcomer leader-member exchange: The contribution of anticipated organizational support," *Journal of Occupational and Organizational Psychology*, 89 (2016): 834–855, https://doi.org/10.1111/joop.12157

62 Tatjana Schnell and Carmen Hoffmann, "ME-work: Development and validation of a modular meaning in work inventory," *Frontiers in Psychology*, 11 (December 2020), https://doi.org/10.3389/fpd/g.2020.599913

63 Christopher D. Nye, James Rounds, Cristina D. Kirkendall, Fritz Drasgow, Oleksandr S. Chernyshenko, and Stephen Stark, "*Adaptive vocational interest diagnostic: Development and initial validation*" (Technical Report 1378) (2019). Fort Belvoir, Virginia: United States Army Research Institute for the Behavioral and Social Sciences.

64 Meyer, Stanley, Herscovitch, and Topolnytsky, *"Affective, continuance, and normative commitment,"* 20–52.

65 Arla Day and Krista D. Randall, "Building a foundation for psychologically healthy workplaces and well-being," In Arla Day, E. Kevin Kelloway, and Joseph J. Hurrell Jr. (Eds.), *Workplace well-being: How to build psychologically healthy workplaces*, pp. 3–26 (2014). West Sussex, UK: Wiley Blackwell Publishing.

66 Stavroula Leka and Aditya Jain, *Health impact of psychosocial hazards at work: An overview*, (2010). Geneva: World Health Organization, http://apps.who.int/iris/bitstream/10665/44428/1/9789241500272_eng.pdf
67 W. Kent Anger, Diane L. Elliot, Todd Bodner, Ryan Olson, Diane S. Rohlman, Donald M. Truxillo, Kerry S. Kuehl, Leslie B. Hammer, and Dede Montgomery, "Effectiveness of total worker health interventions," *Journal of Occupational Health Psychology*, 20, no. 2 (2015), 226–247, https://dx.doi.org/10.1037/a0038340
68 J.M. Conway, "Interfaith capacity building," *Journal of College and Character*, 19, no. 3 (2018): 236–242, https://doi.org/10.1080/2194587X.2018.1481102
69 Lucy Mayblin, Gill Valentine, and Johan Andersson, "In the contact zone: Engineering meaningful encounters across difference through an interfaith project," *The Geographical Journal*, 182, no. 2 (2016): 213–222, https://doi.org/10.1111/geoj.12128
70 Gagné and Deci, *"Self-determination theory,"* 331–362.

Martin I. Jones, Sophie L. Wardle, and Fiona N. Koivula

"A day well spent procures a happy sleep"

And So to Bed: Sleep, Well-being, and Human Performance[1]

Leonardo da Vinci[2]

Anyone who has filled their day with physical and mental labor will confirm that da Vinci's words are valid. A good sleep typically follows a day of exhausting or stimulating activity. These sentiments were echoed also by the legendary Beatles songwriters John Lennon and Paul McCartney, "it's been a hard day's night; I should be sleeping like a log." But, what about the relationship between a good night's sleep and the quality we experience the following day? Does good sleep increase the quality of our wakeful period, and does a poor night's sleep decrease the quality of the awake period? The purpose of this chapter is to answer those questions in the best possible way.

In this chapter, we will examine sleep as a pillar of well-being and human performance. We will describe how sleep is controlled, discuss the evidence that reveals the harmful effects of restricted sleep (and the benefits of quality sleep), and offer evidence-based suggestions to improve sleep. By the end of the chapter, readers should recognize healthy sleep and identify potential ways to maximize sleep.

What Is Sleep?

All living organisms engage in periods of rest or convalescence that resemble sleep.[3] From the simplest single-celled organisms to the most complex of biological species, it seems that everything and everyone sleeps in some shape or form. In humans, sleep is a reversible behavioral state characterized by natural recurrence (usually daily), altered consciousness (and brain wave activity), inhibited sensory activity, decreased ability to react to stimuli, inhibited control of voluntary muscles, and reduced interaction with the environment.[4] In addition, someone sleeping would usually show postural recumbence, motionlessness, and closed eyes. These characteristics are driven by altered brain activity that is coordinated across both hemispheres of the brain (bi-hemispheric sleep), though some species of aquatic mammals and birds are capable of isolating these characteristics to one hemisphere (uni-hemispheric sleep), to maintain vital functions or remain alert to prey.[5]

The altered brain waves can be measured using electroencephalography (EEG) and generally show that sleep can be separated into two main types: Rapid eye movement (REM) sleep and non-rapid eye movement (NREM) sleep.[6] REM and NREM sleep occurs in cycles of approximately 90-minutes throughout the night and are associated with distinct patterns of brain activity, eye movement, and capacity for muscle activation (i.e., REM sleep is characterized by muscle paralysis). The ratio of NREM to REM sleep in each 90-minute period changes throughout the night, with more NREM sleep in the first half of a sleep episode (i.e., the initial sleep cycles) and greater proportions of REM sleep in later 90-minute cycles.[7] The ratio of NREM to REM sleep also changes following periods of sleep loss, with more NREM (precisely stage 3) sleep recorded after total sleep deprivation, compared with a good night's sleep. The distribution of sleep stages is also known as *sleep architecture*.[8]

NREM sleep can be subdivided into three subordinate sleep stages that correspond with an increasing depth of sleep (Stage1 to Stage3). Stage 1 sleep occurs mainly at the beginning of each sleep cycle and is characterized by theta brain waves and slow eye movement.[9] It represents approximately 5 % of the time spent asleep (assuming a normal sleep duration of 7 to 9 hours).[10] Stage 2 sleep involves no eye movement and dreaming in this stage is very rare.[11] Theta brain waves also characterize stage 2 sleep but with the addition of spikes in activity called spindles and K-complexes.[12] Healthy people (i.e., people with no sleep disorders) will spend around 45 % of the night in stage 2 sleep.[13] Stage 3 sleep is characterized by deep, slow-wave sleep. Dreaming is more common in this stage than in other NREM sleep stages, though not as common as in REM sleep. Stage 3 is characterized by low frequency but high amplitude delta and theta brain waves.[14] Healthy people will spend around 25 % of the night in stage 3 NREM. REM sleep, which represents approximately 25 % of the night, involves low frequency and low amplitude beta waves (similar to beta and alpha waves observed when awake).[15] By definition, REM sleep is associated with rapid eye movements and with dreaming. This stage of sleep is also associated with higher breathing rate, heart rate, and blood pressure, compared with NREM stages and an inhibition of skeletal muscles called atonia that may serve to stop people from living out their dreams.[16]

How Is Sleep Regulated?

Before we can determine what is negatively influencing sleep or strategize how to fix or improve it, we need to understand how sleep is controlled. This section involves a brief description of the physiological basis of sleep to help readers understand what is going on "under the hood" while asleep. By understanding how sleep is controlled readers can better appreciate when (and where) specific sleep countermeasures can be used and the mechanisms of effect. Readers can choose to skip this section without impeding their comprehension of the rest of the chapter; but it is important to understand that sleep is an active, organized, and highly a complex state that involves all of the brain's neurotransmitters systems. The analogy of sleep being a period of "down time", like switching a computer off at the end of the night, could not be further from the truth.

Multiple brain regions generate and contribute to sleep and wakefulness.[17] Therefore, it is essential to acknowledge that no single brain region initiates or controls sleep. Rather numerous brain areas are implicated in sleep control. Two neural processes regulate when we sleep and wake: Process C (circadian rhythm) and Process S (sleep drive). Circadian rhythms influence the time of day in which sleep is likely to occur. The sleep drive provides a cumulative "duration of time awake" input to regulate sleep onset.

Process C. Most people have typical patterns of sleep (i.e., sleep during the night and wake during the day) that closely match a biological endogenous oscillation process of approximately 24 hours. Process C is driven by a circadian (*Circa*: around and *dian*: a derivative of the Latin Diem for the day) clock known as the circadian rhythm. The circadian rhythm is reflected in changes in body temperature, sleep tendency, hormone secretions, and cognitive and physical performance. For example, body temperature peaks during the late afternoon to early evening and then falls rapidly toward the early morning.

Similarly, muscle strength peaks in the early evening, and muscle stiffness (i.e., resistance to motion) of the knee joint is lowest in the early evening.[18] Researchers have also revealed circadian variation in psychomotor vigilance, with the lowest levels during the nighttime. Problems associated with insufficient sleep occur at predictable times of the day according to the circadian rhythm. They are most noticeable during the second 12-hour cycle, especially within the 3 hours between 13:00 and 16:00. Equally, misalignment of circadian rhythms and sleep-wake rhythm leads to profound neurobehavioral decrements, becoming cumulative over time. These misalignments are apparent to anyone who has suffered from jetlag or worked night shifts.

Sleep and wakefulness are linked to exogenous cues such as light (i.e., day and nighttime) but are controlled endogenously. Light level is "sampled" by the suprachiasmatic nuclei, a neural structure of approximately 50,000 neurons (i.e., a tiny brain structure) that sits above the chiasm of the optic nerve in the anterior hypothalamus and acts as a master clock. Neurons in the ventrolateral suprachiasmatic nuclei have the ability for light-induced gene expression, which permits synchronization with the outside world (i.e., sleep during darkness and wakefulness during light). Neurons in the dorsomedial suprachiasmatic nuclei have an endogenous rhythm such that the circadian rhythm can operate under constant darkness and constant light (e.g., in polar regions, underwater, in areas of light pollution, or underground). The suprachiasmatic nuclei projects into other areas of the hypothalamus and thalamus to regulate things like body temperature and the Hypothalamic–Pituitary–Adrenal (HPA) axis, which is involved in the response to stress. The suprachiasmatic nuclei also projects to the pineal gland, which produces melatonin for generation of sleep. As a master clock, the suprachiasmatic nucleus influences expression of clock proteins throughout the body that run through a biochemical cycle of about 24 hours. The clock proteins influence the daily timing of hormone release and other bodily functions.[19]

S-Process. The second process that regulates sleep is a homeostatic drive for sleep that increases with time awake and dissipates with time asleep. The longer a person is awake, the more they feel a need to sleep. This process is labeled the S-Process. Typically, after around 16 hours of wakefulness, the homeostatic drive for sleep, or sleep pressure, will reach such levels that people feel the need for sleep and consequently obtain an increased duration and intensity of sleep (compared with attempting to sleep during low sleep pressure). This peak in the drive for sleep usually coincides with the dip in wakefulness associated with the circadian sleep process.

Endogenous somnogens (substances that cause or induce sleepiness) build throughout periods of wakefulness. Numerous somnogens influence sleep; however, the most studied is adenosine. Adenosine is a crucial mediator of sleep pressure. Researchers have shown that increasing adenosine levels increases sleep while decreasing adenosine levels increases wakefulness. Being awake increases levels of adenosine initially in the basal forebrain and then throughout the cortex. Adenosine levels then decrease during sleep. Higher levels of adenosine slow down cellular activity and weaken arousal.

Adenosine levels are influenced by neuronal activity and are a secondary byproduct of the dephosphorylation of adenosine triphosphate (ATP), adenosine diphosphate (ADP), adenosine monophosphate (AMP), and cyclic adenosine monophosphate (cAMP). Thus, throughout wakefulness, the volume of adenosine builds through normal cellular processes and increases the need for sleep.

Blatter and Cajochen stated that waking neurobehavioral performance is regulated by a delicately tuned interaction of sleep homeostasis (i.e., time asleep or time awake) and circadian rhythmicity.[20] The circadian rhythm works synergistically with the homeostatic sleep drive to regulate when we feel sleepy and fall asleep. This two-process model results in consolidated periods of approximately 16 hours of wakefulness and about 8 hours of sleep for most people.

In healthy people, the circadian alerting system drops in the evening (around 21:00). At around this time, melatonin secretion begins (the dim light melatonin onset: DLMO). Melatonin is a sleep-promoting hormone released during darkness from the pineal gland in the brain. Thus, melatonin acts like a starting pistol that signals the beginning of sleep. The combination of melatonin release with the high homeostatic sleep drive initiates a cascade of neural processes that decreases alertness and permits sleep onset. During sleep, the homeostatic sleep drive dissipates, but the circadian alerting system stays low (until around 04:00 in healthy individuals), keeping us asleep. The circadian alerting system then builds back up into the early morning, as remaining somnogens are cleared, to initiate wakefulness.

A problem faced by military personnel, or anyone working at night or moving across time zones, is that some operations may be required at times discordant with the circadian rhythm. In addition, because of the interplay of C and S processes, military personnel need to recognize that capacity for performance at 15:00 is not the same as 03:00 even if the individual obtains adequate sleep. Therefore, it is not surprising that accidents, caused in part by human error, occur during times when the circadian alertness system is low.

For example, Belenky and colleagues described an after-action review of personnel involved in a friendly fire incident during Operation Desert Storm (i.e., the Gulf War).[21] They described how, at dusk, after 48-plus hours of continuous operations, a platoon of six U.S. Bradley fighting vehicles set up a screen line. At around 01:00, the U.S. personnel spotted hot spots in the thermal imaging sights moving toward the screen line. The U.S. soldiers engaged the hotspots when they were identified as Iraqi armored personnel carriers. In the ensuing firefight, the U.S. forces destroyed all Iraqi vehicles. However, the aftermath of the firefight also revealed the destruction of two of the U.S. Bradley fighting vehicles. Later it transpired that "friendly fire" was responsible. The crews accountable for destroying the U.S. vehicles had mistaken their compatriots for the enemy. While they believed they were facing forward toward the enemy, they were in fact facing their own people and had enfiladed their line (i.e., fired on friendly forces from the side). The authors concluded that the crews were sleep-deprived based on the self-reported fragmented sleep before the friendly fire incident. As a result, they had lost their orientation and grasp of the tactical situation.

Belenky and colleagues concluded that the crews of the fighting vehicles that fired on their friends held to the sound tactic to fire on anyone approaching their front. However, they were no longer clear as to where the front was.[22] The important point to emphasize is that the consequences of sleep restriction for military personnel is of a different order of magnitude to the consequences for most civilians. At the same time military sleep restriction is often endorsed and praised as a sign of toughness and zeal. The consequences of sleep restriction in military personnel is equivalent to the consequences of civilians working in safety critical industries such as aviation, nuclear power, or the rail industry (i.e., large-scale accidents and death). The main difference is that sleep restriction in the military is sometimes unavoidable for operational reasons and the culture in some military units reinforces the idea of sleep restriction as necessary and needed.

Why Do We Sleep?

While the sleep function is complex and multifaceted, it is apparent that sleep represents a fundamental physiological requirement for human survival. Researchers and practitioners agree that if individuals do not sleep (i.e., total sleep deprivation), they will eventually suffer systemic malfunction and ultimately death (as evidenced in sleep disorders, such as fatal familial insomnia – a rare genetic degenerative brain disorder).[23] Moreover, because sleep plays a pivotal role in many processes, including growth and development, brain waste clearance, modulation of immunity, memory, and emotional regulation, disrupted sleep can contribute to the breakdown of various bio-psycho-social processes.

Sleep disruption (i.e., breaks from the normal unrestricted sleep each day) can be subcategorized into two types: sleep restriction and (total) sleep deprivation. Total sleep deprivation involves periods when an individual has no sleep during a 24-hour day. Therefore, total sleep deprivation is the complete absence of sleep or, put differently, chronic forced wakefulness. Chronic partial sleep deprivation, also known as chronic sleep restriction, is when an individual has at least one week of consecutive days with six or fewer hours of sleep per 24-hour period. Acute partial sleep deprivation is where sleep is less than "normal" (but not totally removed) and its effect can be observed in as little as one day. While there is no recognized cutoff, at which point we can label someone as partially sleep-deprived, it is crucial to acknowledge that any amount of sleep loss can be reflected in a degraded performance and well-being in some people. It is also essential to recognize that there are ranges of individual differences that influence the magnitude of sleep loss that one can manage. Some people are robust to acute partial sleep deprivation and suffer little if no ill effects.

Sleep restriction occurs when humans fall asleep later or wake earlier than usual (i.e., the typical sleep-wake cycle is partially disturbed). Therefore, sleep restriction comprises less than expected but not zero hours of sleep. For example, sleep restriction could include limited sleep during the weekday and an attempt to binge on sleep during weekends. Similarly, sleep restriction could consist of less than the recommended period of 7 to 9 hours per day of sleep for the duration of a military operation (e.g., one week) that typically returns to normal once that phase in operations ceases. Sleep restriction is more prevalent than total sleep deprivation in military settings primarily because of operational requirements such as traveling (i.e., jet lag), shift work (i.e., nighttime operations), and prolonged work hours. In contrast to sleep restriction, where some sleep (albeit limited or broken sleep) occurs, acute total sleep deprivation infrequently occurs outside the sleep laboratory.

Sleep restriction is characterized by the interruption of normal sleep with brief periods of wakefulness during the night. This type of sleep restriction is also known as sleep fragmentation and often interrupts the normal progression and sequencing of sleep stages.[24] As a result, an individual suffering from sleep fragmentation will likely have less time in physiological sleep than the time in bed (i.e., reduced sleep efficiency).

Banks and Dinges coined the term selective sleep stage deprivation, which involves losing specific physiological sleep stages.[25] Selective sleep stage deprivation can occur if sleep fragmentation is isolated to one particular sleep stage. For example, certain medications or medical conditions could restrict or extend REM sleep relative to an individual's regular sleep cycles. Anyone who limits their sleep will have some form of selective sleep restriction. Given that REM sleep increases in duration towards the end of normal sleep, people who have shortened sleep duration (e.g., 5 hours rather than 8 hours) will likely experience a reduction in overall time in REM sleep. Similarly, people who attempt to catch up on sleep after a period of extended sleep debt will probably spend more time in NREM relative to REM sleep, at least in the short term.

Sleep disruption is a pertinent topic for military personnel because many military operations are characteristically irregular and often put military operators into acute and chronic sleep deprivation and sleep restriction. With the need to provide a 24-hour capability comes the need to manage sleep to reduce the adverse effects of sleep restriction on cognitive, emotional, and physical health, well-being, and performance.

Successful management of the sleep-wakefulness cycle contributes to optimal physical and mental health in addition to occupational performance.[26] Evidence from various sources suggests that transient sleep disruption, specifically sleeping less than the recommended 7 to 9 hours per night, reduces many higher functions such as memory, abstract reasoning, decision-making, and attention.[27] In addition, Killgore suggested that inadequate sleep impairs effective communication, sensitivity to risk, and creative, innovative thinking, mainly when emotion processing networks are involved.[28]

To date, there has been a lack of consensus relating to the effect of sleep disruption on physical performance. Aerobic endurance performance appears to be consistently compromised by acute sleep restriction and sleep deprivation. However, anaerobic performance and muscle strength and power seem to be less affected by sleep loss, albeit not consistently. In military settings, physical performance, particularly aerobic capacity, muscle endurance, and military-specific physical activity, is impaired by sleep loss.[29] Still, the independent effects of sleep loss cannot be easily differentiated from other aspects of military activity affecting physical performance, including high energy expenditure, and limited dietary intake. In general, there appears to be a less marked decrease in physical performance compared with cognitive performance.[30] However, changes in cognitive performance could mediate physical performance degradation. For example, Van Cutsem and colleagues suggested that inadequate sleep influences perceived effort during endurance tasks (e.g., tasks feel harder), which subsequently degrades physical performance.[31]

Shattuck and colleagues classified the consequences of poor sleep practices based on the duration of effects (i.e., short term, intermediate term, and long-term effects).[32] Short term effects of poor sleep practices included degraded reaction time, increased risk of accident, decreased vigilance, inconsistent logical reasoning, reduced short term memory, increased negative mood, increased risk of injury, increased depression, anxiety, and paranoia symptoms and increase in stress hormone production (i.e., cortisol).[33] Intermediate term risks included loss of motivation, reduced memory, longer time needed to train, decreased immunity, caffeine addiction, and failure to accomplish missions.[34] Long term effects included circadian scarring (i.e., sleep disturbances and elevated daytime sleepiness after years of shiftwork), metabolic disorders, and chronic disease (e.g., dementia, Alzheimer's disease, some cancers, type II diabetes, and cardiovascular disease).[35] Despite these risks, many people are chronically sleep restricted, and some even glorify sleep disruption. The effects of sleep disruption should not be trivialized and indeed should not be celebrated as a badge of honor, toughness, or zeal.

Are We Getting Enough Sleep?

Most of us spend around one-third of our lives sleeping; however, approximately 35 to 40 % of the U.S. adult population report problems with falling asleep or daytime sleepiness.[36] Yong Liu and colleagues also reported that approximately one third of U.S. adults surveyed in 2014 reported averaging less than 7 hours sleep per night.[37] Compared with civilian data, it appears that short duration sleep is equally prevalent in service personnel.[38] Meadows and colleagues revealed that around two thirds of U.S. military personnel reported averaging less than 8 hours sleep per night, however some services (i.e., Marine Corps) reported significantly more short duration sleep than others (i.e., Air Force and Coast Guard).[39]

LoPresti and colleagues described self-reported sleep duration of between 5 and 6 hours per day in deployed U.S. Army personnel in a military population.[40] Importantly, their results also revealed that 15 % of their sample described accidents that affect the mission, 50 % of which were attributed to inadequate sleep. These statistics are of particular concern because military personnel are often in safety and mission critical roles and accidents can therefore have catastrophic (e.g., fatal) consequences.

Troxel and colleagues also examined whether military personnel experienced sleep disruptions.[41] The results of a study investigating sleep in a sample of 1851 U.S. military personnel across all branches of the U.S. military (excluding the Naval reserve) also showed a high incidence of sleep problems. Around one-third of the sample reported fewer than 5 hours of sleep per night and were categorized as extreme short sleepers. In addition, approximately 50 % of the same sample reported clinically significant sleep problems based on the Pittsburgh Sleep Quality Index (PSQI).[42] Interestingly, these sleep problems did not appear to be the result of deployments or combat stress but rather indicated general sleep issues in a large proportion of the U.S. military. Working nights, military shift patterns, rapid time zone changes, sleeping in austere environments, and operational tempo are all likely antecedents of poor sleep within military personnel. Assuming that these results represent other militaries, it appears evident that a significant proportion of soldiers, sailors, aviators, and marines need help to manage their sleep-wakefulness cycle. Sleep problems are a significant issue for military personnel because not only is lack of sleep associated with aforementioned ill health, but it can also influence the ability to assess and appropriately respond to threats in training and on deployment.

Van Dongen and colleagues conducted a chronic sleep restriction experiment where participants were randomized into one of three sleep durations (4 hours,

6 hours, or 8 hours' time in bed per night), which were maintained for 14 nights followed by three nights of recovery.[43] In addition, one group of participants were randomized in a total sleep deprivation group (i.e., 0 hour time in bed per night) that last for three nights. The researchers administered the psychomotor vigilance task (PVT) and a working memory test (digit symbol substitution task) to the participants every 2 hours during scheduled wakefulness. Additionally, participants were invited to complete measures of subjective (self-rated) sleepiness. The results revealed that chronically restricting sleep to 4 hours and 6 hours per night progressively eroded psychomotor vigilance and working memory. The authors stated that the results of this experiment provided convergent evidence for the adverse effects of chronic sleep restriction on cognitive functions.[44] An additional finding, regarding the subjective sleepiness scores, suggested that once sleep restriction is chronic, those chronically sleep restricted participants cannot reliably gauge their actual sleepiness levels. Equally, the results may reveal that, as long as people are receiving approximately 4 hours of sleep each night, they do not experience a sense of sleepiness anywhere near the levels found for total sleep deprivation. These results may shed light on the phenomenon of perceived adaptation to sleep restriction. The results from this study revealed that restricted sleepers do not feel sleepy (compared with total sleep deprivation) but their cognitive functions are impaired. There was no evidence that people adapted to restricted sleep over 14 days, but their perceived sleepiness did begin to plateau.[45] Consequently, people might report feeling "fine" after periods of restricted sleep duration (subjective sleep) but despite this perception, they are significantly impaired. Similarly, people may believe that they can train themselves to need less sleep but the same pattern of subjective sleepiness, but impaired cognitive performance will likely emerge.

Unfortunately, sleep deprivation is profitable. Many people have made money, and continue to make money, off the supposed sleep deprivation epidemic, fueled by social media influencers, by overstating and magnifying the risk and creating and promoting a range of tools, technologies, pharmaceuticals, and supplements that "cure" or treat sleep deprivation. It is likely that the working requirements of the military can create or exacerbate sleep problems and therefore military personnel seek hyped up "cures" before seeking medical advice. While we believe that sleep disruption is prevalent in military populations at a rate similar to (or possibly higher than) civilian samples, we do not wish to create anxiety in readers by endorsing an oversimplified narrative that links sleep loss to morbidity.

It is essential to recognize that several individual differences (e.g., genes, hormones, environment, previous sleep, and sleep loss) influence sleep duration and sleep architecture. Therefore, the range of adequate sleep duration is relatively broad, and most public health recommendations are based on a population average. The golden number of 8 hours per night (or a range of 7 to 9 hours) is a rule of thumb, and some people exist toward the tails of the distribution of sleep duration. For most of the population, 7 hours to 9 hours per night will be optimal, but this number is a guide and not a rule. Some people will only require 5 hours and others 11 hours, although these people are likely to be in the minority.

In addition to between-person differences in sleep need, evidence from the observation of pre-industrial societies (i.e., modern-day hunter-gatherer tribes in Africa and South America) revealed variation in day-to-day sleep.[46] Hunter-gatherers and horticulturalists sleep for an average of 6.4 hours per day with a range of 5.7 to 7.1 hours, which would be considered at the lower end of the sleep advice for industrialized societies. The pre-industrialized sample slept an additional 1 hour per day during the winter compared with summer, and daily changes could influence variability in sleep duration at ambient temperature. Yetish and colleagues stated that the onset of sleep varied daily but was typically around 3 hours after sunset and usually coincided with a drop in ambient temperature.[47] The sleep period consistently occurred during the nighttime period of falling environmental temperature (although napping increased during summer months), was not interrupted by extended periods of wakefulness, and terminated near the nadir of daily ambient temperature. The daily cycle of temperature change could be a potent natural regulator of sleep; however, living indoors with predictable heat sources (and light) could have blunted our natural variation in daily sleep duration.

The key message is that sleep varies, and if an individual falls outside of the 7 hours to 9 hours per day range, or if sleep duration wavers, it is not necessarily a cause for concern. If changes in sleep duration, or sleep behavior, occur alongside significant daytime sleepiness or changes in health and well-being, it could be worthwhile consulting a medical professional.

In sum, there is extensive individual variation in sleep duration and timing.[48] Consequently, it is vital to define sleep needs based on responses to a set of sleep criteria. The main symptom of sleep loss is excessive daytime sleepiness. Other important information to consider is whether individuals are dependent upon an alarm clock or another person to get out of bed, whether they sleep extensively (get up late) on free days or take a long time to wake up and feel alert.[49] Additionally, other signs of inadequate sleep include feeling sleepy and irritable during the day, feeling the need for a mid-afternoon nap to function adequately, and inability to concentrate and exhibiting overly impulsive behaviors.[50] Finally, craving caffeinated and sugar-rich drinks or experiencing increased worry, anxiety, mood swings, and depression could signpost the need for more sleep.[51]

Promoting Sleep and Sleep Countermeasures

In the final section of this chapter, we will consider things that people can do during wakefulness to influence the quality and quantity of subsequent sleep. We will also consider sleep countermeasures that people can adopt when everyday sleep routines are impossible (e.g., high operational tempo, night working, shift working). We define a sleep countermeasure as a measure or action taken to counter or offset the effects of prolonged wakefulness or restricted sleep (e.g., sleepiness, performance degradation). First, an essential note of caution: countermeasures that offset sleep-related cognitive and physical performance deficits are viable, but it is necessary to recognize that these countermeasures all have limited returns and cannot replace actual sleep. Therefore, sleep countermeasures should not be used to replace sleep but instead be used as a last resort when opportunities for sleep are not available.

In this final section, we will intentionally avoid discussion of sleep medications (e.g., Ramelteon, Zaleplon, Zolpidem+, Eszopiclone, Alprazolam, Lorazepam, Trazodone, Mirtazapine, Suvorexant, Melatonin) and alertness drugs (e.g., Modafinil, amphetamines) that typically (dependent on a country's legislation) require prescription and administration by a medical professional. Instead, we will focus on the "therapies of wakefulness" that mainly do not require medical supervision and are primarily under the control of the individual. These therapies of wakefulness include sleep hygiene, sleep extension, napping, manipulation of light, and caffeine use.

Sleep hygiene. When people engage in non-sleep-promoting activities such as intense exercise, use of alcohol, or stimulants (e.g., caffeinated beverages), or bathing in bright light before bedtime, physiological responses (e.g., delayed dim light melatonin onset and adenosine receptor antagonism) can negatively influence the onset of sleep (and sleep architecture). Engaging in arousing activities like watching an encapsulating film or reading an enthralling book could also affect sleep through cognitive mechanisms (e.g., motivational processes and increased alertness). Because the aforementioned non-sleep-promoting behaviors are amenable to change, sleep hygiene interventions focus on using education and training to modify maladaptive behavior. Educational strategies that commonly appear in cognitive behavioral therapy for insomnia (CBTi) could be used by normal sleepers (i.e., those without insomnia) to help promote improvements in sleep, particularly if chronically sleep-restricted by choice through the delayed onset of sleep. These educational sleep hygiene strategies aim to improve sleep by addressing modifiable factors that may disturb sleep, such as light, noise, and the ingestion of caffeinated or alcoholic drinks. It is not within the scope of this chapter to review all the cognitive and behavioral techniques that CBTi practitioners use therefore interested readers should consult a relevant review.[52]

We contend that there is no need to evoke changes in all pre bedtime behaviors in relatively healthy sleepers. However, reflecting upon current non-sleep-promoting activities and considering whether adopting sleep-promoting behaviors could improve sleep (and subsequently well-being and performance) is likely beneficial for most people.

Foster described a range of individual actions that could help to achieve better sleep.[53] These actions were divided into activities that can be undertaken during the day, before bed, in the bedroom (or other sleeping space), and in bed to help improve sleep. During the day, Foster recommended maintaining a sleep and wakefulness routine where people go to bed and get up at approximately the same time each day, including weekends. He recommended getting as much natural morning light as possible (or if natural light is not feasible, the timed use of lightboxes can help regulate sleep).[54] He stated that naps should last no longer than 20 minutes and not within 6 hours of bedtime. Exercise and eating food should also be avoided close to bedtime (due to increased core body temperature accompanying vigorous physical activity and eating). Caffeine-rich foods and drinks (e.g., coffee), and other stimulants such as nicotine, should be monitored to avoid overuse (i.e., >450 mg caffeine per day) and ideally avoided up to 10 hours before bedtime (depending on individual variability in response to caffeine).

Before bedtime, Foster recommended reducing light levels and stopping using electronic devices approximately 30 minutes before bedtime.[55] In addition, Foster recommended the avoidance of prescription sedatives, alcohol, antihistamines, or other people's sedatives as a means to induce sleep. Finally, he suggested finding a pre-bedtime strategy that works best for the individual. This routine could include the aforementioned pre-bedtime behavior in addition to stress-relieving activities such as listening to music, reading, mindful meditation, or progressive muscle relaxation.[56]

In the bedroom, Foster endorsed a sleeping environment comprised of a moderate temperature (e.g., 18°C), low noise (e.g., low-level white noise), or silence, darkness, which is also devoid of personal electronic devices and televisions.[57] In addition, he suggested that people prone to clock watching should consider either removing the timepiece from the sleep environment or moving it somewhere that cannot be seen (particularly if the clock is luminescent). Finally, he warned over obsessing over sleep apps (which could include wearable devices) that can provide (sometimes erroneous) feedback on sleep metrics, including, but not limited to, sleep duration, sleep architecture, and sleep efficiency. Baron and colleagues coined the term "orthosomnia" to reflect the condition where people seek treatment for self-diagnosed sleep disturbances such as insufficient sleep duration and insomnia due to feedback from their sleep trackers and apps.[58] Obsessive perfectionistic striving to obtain ideal sleep as a method of optimizing daytime functioning is ironically more likely to disrupt sleep than facilitate sleep because such obsessions can reinforce sleep-related anxiety, particularly if those obsessive people are unable to hit their arbitrary sleep metrics.

In the final category of individual actions to achieve better sleep, Foster addressed the bed itself.[59] He recommended sleeping in a sufficiently large, comfortable, and relaxing bed (including bed linens and pillows). For people sharing beds, if one's bedfellow is disrupting sleep through snoring, particular bedtime habits (e.g., reading under light), a preference for a different perception of what constitutes comfort (e.g., mattress softness) or by monopolizing the space permitted for sleep, it might be helpful to sleep in separate beds or separate rooms. Failing that, devices such as eye masks, earplugs, or mattress toppers might be beneficial in some sleep environments.

Alcohol. The concept of a nightcap to facilitate sleep is common in some cultures and many insomniacs self-medicate with alcohol to help them fall asleep.[60] While alcohol may induce sleep, the effect of alcohol wears off with continued use and therefore the volume of alcohol often increases in response. The consumption of alcohol to increase sleep onset has two main effects. Firstly, the volume of liquid consumed can increase sleep disturbance because of nocturia, the condition that causes one to wake up during the night to urinate. Secondly, alcohol interrupts normal sleep architecture. Lockley and Foster reported that during periods of heavy alcohol use (and up to two years of abstinence) there are disruptions to the volume of slow waves sleep states, suppressed REM sleep, fragmented sleep, and reduced sleep duration.[61] The sedative effect of a nightcap might help induce sleep but the consequences of alcohol on subsequent sleep architecture means that alcohol should not be endorsed as a sleep aid.

Sleep extension. Allowing an extended period of sleep, in anticipation of future sleep restriction, can alleviate the slope of performance decline over prolonged wakefulness. This idea of increasing the time spent in bed to allow people to expand their total sleep time is reflected in the Leader's Guide to Soldier Health and Fitness developed by the Office of the U.S. Army Surgeon General.[62] In this leader's guide, there is the broad recommendation that a soldier sleeps at least 7 hours per night whenever possible but recommends at least 9 hours per night in preparation for episodes of inadequate sleep. Individuals can extend their sleep in two ways. Firstly, extending the duration of nocturnal sleep (e.g., extending time in bed at the end of the day) or secondly, by napping during the day to increase overall time asleep during a given 24-hour cycle.

Arnal and colleagues investigated the effects of six nights of sleep extension on sustained attention and sleep pressure before and during total sleep deprivation and after subsequent recovery sleep.[63] Results revealed that extending sleep (9.8 hours in bed) vs. habitual sleep (8.2 hours in bed for the control group) for six days before 22 hours of sleep deprivation improved sustained attention (both fewer lapses and faster speed on the psychomotor vigilance task: PVT). Furthermore, sleep extension limited the increase of PVT lapses and microsleeps during total sleep deprivation without changing PVT speed. Crucially, the positive effects of sleep extension were still present after one night of recovery sleep.[64]

However, these results and the reason that sleep extension improves (or maintains) performance and well-being are contentious. Some researchers believe that sleep can be banked; conversely, others suggest that sleep extension just clears a previous sleep debt (i.e., it only works if you are recently sleep deprived). If an individual is in a state of sleep restriction and has built up a sleep debt, giving them an extended time in bed can reverse that sleep debt. Effectively, reducing sleep debt represents a regression to baseline performance levels (when people are adequately slept/rested). It is less clear whether extending sleep in "healthy" people will reveal similar positive effects. If an individual does not have a sleep debt, it is unlikely that extending sleep will improve performance.

Napping. By sleeping for short durations during the daytime before periods of extended wakefulness, people can add to their overall time asleep in a 24-hour cycle, and evidence suggests that by napping, they can reduce the building sleep pressure and reverse degradation in performance. The optimum nap duration must balance the improvements in performance following waking and the adverse effects of sleep inertia. Sleep inertia is a state of impaired sensory-motor and cognitive performance that is present immediately after awakening. Typically, the grogginess associated with sleep inertia will dissipate after around 30 minutes of waking. Napping can lead to sleep inertia if the nap encroaches on deeper stages of sleep. Longer naps (i.e., more than 90 minutes) will likely have a more significant recuperative effect; however, there is also a risk of waking during slow-wave sleep and experiencing sleep inertia. To benefit from the positive effects of napping and reduce the risk of sleep inertia, the nap duration can be reduced to 10 to 30 minutes (i.e., staying in the lighter sleep stages). Moreover, consuming caffeine before, say, a 30-minute nap can increase alertness upon waking because most caffeine products take time (e.g., 30 minutes) to influence alertness.

The effectiveness of naps to improve cognitive performance in sleep-restricted individuals is well established. Researchers have shown that naps improve cognitive functioning (e.g., short-term memory, alertness, psychomotor, performance), mood, and subjective and objective sleepiness.[65] Souabni and colleagues evaluated the effectiveness of daytime napping opportunities on athletes' physical and cognitive performance by reviewing existing research.[66] The results of their review revealed that napping improved endurance performance, short-term physical performance, and specific skills performance. In addition, napping improved reaction time, attention, and short-term memory.

Furthermore, Souabni and colleagues revealed that more extended nap opportunities (e.g., 90 min) improved physical and cognitive performance and lowered fatigue.[67] In addition to daily napping, naps can be taken during the night during night shifts. Dutheil and colleagues reviewed the benefits of napping during night shift working on cognitive performance.[68] Based on the results of 18 studies and 494 participants, they concluded that napping during night shifts also improves cognitive performances, especially attention.

Manipulation of light. Light is a powerful "zeitgeber" that entrains the circadian rhythm to a 24-hour cycle. Because of the evolution of neural systems that respond to light (e.g., sunlight), it is possible to use artificial sources of light to subvert these systems. Intentional exposure to light and the avoidance of light can therefore be used to sustain alertness and promote sleep.

In the sleep hygiene section of this chapter, we highlighted Foster's recommendation to get as much natural morning light as possible (or use lightboxes) to help synchronize the circadian rhythm.[69] However, light can have other effects on performance and well-being. Visible light is the portion of the electromagnetic spectrum between 400 and 700 nm. The wavelength of the light in nanometers (nm) governs the color of the light spanning from violet to red. The light-sensitive photopigment, melanopsin, present in photosensitive retinal ganglion cells of the retina, is sensitive to the blue light wavelength (~480 nm) that occurs in natural and artificial light. Thus, a light's impact depends on the degree to which a photo pigment is stimulated, and the magnitude of the signal transmitted to the brain (i.e., the suprachiasmatic nuclei) via the retinohypothalamic tract.

To facilitate sleep onset, it is therefore essential to reduce (blue) light exposure (daylight and artificial light) before bedtime because of the suppressive effects that light exposure has on melatonin production. Blue light can also be used to enhance alertness and other cognitive functions. Chellappa and colleagues examined the effect of blue light, specifically a compact fluorescent lamp that provided 6500 Kelvin (K – a measure of color temperature) and a light output of between 420 and 520nm, on subjective and objective alertness levels when compared with "warmer" lower intensity lights (i.e., 2500K and 3000K).[70] Results revealed that the 6500K light source was more effective than light at 2500K and 3000K in reducing subjective sleepiness and enhancing cognitive performance (i.e., the PVT). Lights at 6500K also significantly attenuated melatonin secretion, reinforcing the previous statement that high-intensity blue light should be avoided close to bedtime.[71]

Use of caffeine. Caffeine is a commonly consumed psychoactive substance that can increase alertness by blocking adenosine receptors in the brain. Caffeine is present in a range of products, including (but not limited to) coffee, tea, chocolate, energy drinks, some medications, fortified chewing gum, and some exercise supplements. The volume of caffeine varies considerably based on the source, brand, and preparation method (e.g., brew time). Decaffeinated beverages contain smaller amounts of caffeine than their caffeinated counterparts do, but they should not be considered caffeine-free. In addition, different sources of caffeine have different onsets of action. For example, caffeine chewing gum has a rapid onset of action of 5–10 minutes, whereas caffeine in drinks and foods has an onset of action of 30–45 minutes. Consuming caffeine is an effective method to sustain performance when working long hours; however, caffeine should not be considered as a replacement for sleep if opportunities for sleep exist.

The effect of caffeine varies between individuals and depends on habitual intake and other factors. Therefore, caffeine consumption differs widely among individuals. In addition, the results of caffeine consumption vary significantly between individuals depending on factors such as volume of daily consumption, age, mass, health status, medication, tobacco use, and genetics. Therefore, individuals who do not usually consume caffeine are likely to feel the effects of caffeine at lower levels of consumption. Similarly, those of smaller body sizes and those with a heightened sensitivity to caffeine could consider testing their caffeine response by trying lower doses (e.g., less than 100 mg) to avoid adverse reactions. If adverse reactions to caffeine intake occur, such as gastric upset, anxiety, or sleep disturbance, it is advisable to discontinue use.

Most people will consume between 200 and 400 mg of caffeine per day, but high habitual caffeine consumers will become tolerant to the alerting effects. It is generally agreed that daily consumption of up to 400 mg of caffeine does not present a health risk for healthy adults. However, sustained caffeine use may increase the risk of dehydration in some people, so fluid and food intake should be monitored and caffeinated, or stimulant drinks are not recommended for rehydration. Caffeine in doses up to 400 mg per day improves cognitive performance (e.g., vigilance, mood, reaction time, perceived effort) in healthy adults. Limiting caffeine intake to less than 400 mg in those with high frequent caffeine consumption could adversely affect performance and well-being and result in withdrawal symptoms (e.g., headache). It is difficult to provide definitive guidelines concerning a minimum and maximum caffeine dose. Still, it is advised to restrict caffeine intake to less than 400 mg per day for most people. Overuse of caffeine can diminish effects on alertness as tolerance to the effects of caffeine develops.

To prevent caffeine from interfering with sleep, it should ideally be consumed within the first 8 to 10 hours since waking. The final dose of caffeine should be taken at most 5 hours before bedtime because caffeine can affect alertness for around 3 hours to 5 hours (dose and individual dependent). Therefore, individuals who consume caffeine late in the day or close to bedtime may experience disturbed sleep.

Conclusion

The purpose of this chapter was to help readers to recognize healthy sleep and identify ways to maximize sleep. Sleep is an essential pillar of well-being and performance. Failure to get regular adequate sleep interrupts many biological systems and can create a cascade of cognitive and physiological effects. Once people understand how their sleep is regulated, they can understand how various behaviors might influence the quality (and quantity of sleep). Many of these daytime behaviors are modifiable, and therefore people can take an active role in their sleep quality.

Notes

1 ©Crown copyright (2021), Dstl. This material is licensed under the terms of the Open Government License except where otherwise stated. To view this license, visit http://www.nationalarchives.gov.uk/doc/open-government-licence/version/3 or write to the Information Policy Team, The National Archives, Kew, London TW9 4DU, or email: psi@nationalarchives.gov.uk
2 Jean Paul Richter, "*The Notebooks of Leonardo Da Vinci*". Available at http://www.gutenberg.org (1888).
3 Chiara Cirelli and Giulio Tononi, "Is sleep essential?" *PloS Biology*, 6 (August, 2008): e216, https://doi.org/10.1371/journal.pbio.0060216
4 Mary A Carskadon and William C. Dement, "Normal human sleep: An overview." *Principles and Practice of Sleep Medicine*, 4, no. 1 (2005): 13–23.

5 Chiara Cirelli and Giulio Tononi, *"Is sleep essential?"*
6 Bryan Kolb and Ian Q. Whishaw. *Fundamentals of human neuropsychology* (2009). New York: Madison.
7 Carskadon and Dement, *"Normal human sleep: An overview."*
8 Ibid.
9 Ibid.
10 Ibid.
11 Ibid.
12 Ibid.
13 Ibid.
14 Ibid.
15 Ibid.
16 Ibid.
17 Ibid.
18 Barry Drust, Jim Waterhouse, Greg Atkinson, Ben Edwards, and Tom Reilly, "Circadian rhythms in sports performance – an update." *Chronobiology International*, 22, no. 1 (2005): 21–44.
19 Steven W. Lockley & Russell G. Foster. *Sleep: A very short introduction* (2012). Oxford: Oxford University Press.
20 Katharina Blatter and Christian Cajochen, "Circadian rhythms in cognitive performance: methodological constraints, protocols, theoretical underpinnings." *Physiology & Behavior*, 90, no. 2–3 (2007): 196–208.
21 Gregory Belenky, David M. Penetar, David Thorne, Kathryn Popp, John Leu, Maria Thomas, Helen Sing, Thomas Balkin, Nancy Wesensten, and Daniel Redmond, "The effects of sleep deprivation on performance during continuous combat operations." *Food Components to Enhance Performance* (1994): 127–135.
22 Gregory Belenky et al., *"The effects of sleep deprivation on performance during continuous combat operations."*
23 Lev G. Goldfarb, Robert B. Petersen, Massimo Tabaton, Paul Brown, Andréa C. LeBlanc, Pasquale Montagna, Pietro Cortelli et al., "Fatal familial insomnia and familial Creutzfeldt-Jakob disease: Disease phenotype determined by a DNA polymorphism." *Science*, 258, no. 5083 (1992): 806–808.
24 Siobhan Banks and David F. Dinges, "Behavioral and physiological consequences of sleep restriction." *Journal of Clinical Sleep Medicine*, 3, no. 5 (2007): 519–528.
25 Ibid.
26 Charles A Czeisler, "Duration, timing and quality of sleep are each vital for health, performance and safety." *Sleep Health: Journal of the National Sleep Foundation*, 1, no. 1 (2015): 5–8.
27 Julian Lim and David F. Dinges, "A meta-analysis of the impact of short-term sleep deprivation on cognitive variables." *Psychological Bulletin*, 136, no. 3 (2010): 375.

28 William D. Killgore, "Effects of sleep deprivation on cognition." *Progress in Brain Research*, 185 (2010): 105–129.
29 Clementine Grandou, Lee Wallace, Hugh Fullagar, Rob Duffield, and Simon Burley, "The effects of sleep loss on military physical performance." *Sports Medicine*, 49, no. 8 (2019): 1159–1172.
30 Hugh Fullagar, Sabrina Skorski, Rob Duffield, Daniel Hammes, Aaron J. Coutts, and Tim Meyer, "Sleep and athletic performance: The effects of sleep loss on exercise performance, and physiological and cognitive responses to exercise." *Sports Medicine*, 45, no. 2 (2015): 161–186.
31 Jeroen Van Cutsem, Samuele Marcora, Kevin De Pauw, Stephen Bailey, Romain Meeusen, and Bart Roelands, "The effects of mental fatigue on physical performance: a systematic review." *Sports Medicine*, 47, no. 8 (2017): 1569–1588.
32 Nita Lewis Shattuck, Panagiotis Matsangas, Vincent Mysliwiec, and Jennifer L. Creamer, "The role of sleep in human performance and well-being." In Michael D. Matthews and David M Schnyer (Eds.), *Human performance optimization*, pp. 200–233 (2020). Oxford: Oxford University Press.
33 Ibid.
34 Ibid.
35 Ibid.
36 Justin S. Campbell, Rachel Markwald, Evan D. Chinoy, Anne Germain, Emily Grieser, Ingrid Lim, and Stephen V. Bowles, "A sleep primer for military psychologists." In Stephen Bowles and Paul Bartone (Eds.), *Handbook of military psychology*, pp. 239–258 (2017). London: Springer.
37 Yong Liu, Anne G. Wheaton, Daniel P. Chapman, Timothy J. Cunningham, Hua Lu, and Janet B. Croft, "Prevalence of healthy sleep duration among adults – United States, 2014." *Morbidity and Mortality Weekly Report*, 65, no. 6 (2016): 137–141.
38 Sarah O. Meadows, Charles C. Engel, Rebecca L. Collins, Robin L. Beckman, Joshua Breslau, Erika Litvin Bloom, Michael Stephen Dunbar, Marylou Gilbert, David Grant, Jennifer Hawes-Dawson, Stephanie Brooks Holliday, Sarah MacCarthy, Eric R. Pedersen, Michael W. Robbins, Adam J. Rose, Jamie Ryan, Terry L. Schell, and Molly M. Simmons, *2018 Department of Defense Health Related Behaviors Survey (HRBS): Results for the active component* (2021). Santa Monica, CA: RAND Corporation, https://www.rand.org/pubs/research_reports/RR4222.html
39 Ibid.
40 Matthew L. LoPresti, James A. Anderson, Kristin N. Saboe, Dennis L. McGurk, Thomas J. Balkin, and Maurice L. Sipos, "The impact of insufficient sleep on combat mission performance." *Military Behavioral Health*, 4, no. 4 (2016): 356–363.

41 Wendy M. Troxel, Regina A. Shih, Eric R. Pedersen, Lily Geyer, Michael P. Fisher, Beth Ann Griffin, Ann C. Haas, Jeremy Kurz, and Paul S. Steinberg, "Sleep in the military: Promoting healthy sleep among US service members." *Rand Health Quarterly*, 5, no. 2 (2015).
42 Daniel J Buysse, Charles F. Reynolds III, Timothy H. Monk, Susan R. Berman, and David J. Kupfer, "The Pittsburgh sleep quality index: A new instrument for psychiatric practice and research." *Psychiatry Research*, 28, no. 2 (1989): 193–213.
43 Hans Van Dongen, Greg Maislin, Janet M. Mullington, and David F. Dinges, "The cumulative cost of additional wakefulness: Dose-response effects on neurobehavioral functions and sleep physiology from chronic sleep restriction and total sleep deprivation." *Sleep*, 26, no. 2 (2003): 117–126.
44 Wendy M. Troxel et al., *"Sleep in the military: Promoting healthy sleep among US service members."*
45 Ibid.
46 Gandhi Yetish, Hillard Kaplan, Michael Gurven, Brian Wood, Herman Pontzer, Paul R. Manger, Charles Wilson, Ronald McGregor, and Jerome M. Siegel, "Natural sleep and its seasonal variations in three pre-industrial societies." *Current Biology*, 25, no. 21 (2015): 2862–2868.
47 Ibid
48 Russell G. Foster, "Sleep, circadian rhythms, and health." *Interface Focus*, 10, no. 3 (2020): 20190098.
49 Ibid.
50 Ibid.
51 Ibid.
52 Vanessa Herbert, Simon Kyle, & Daniel Pratt, "Does cognitive behavioural therapy for insomnia improve cognitive performance? A systematic review and narrative synthesis." *Sleep Medicine Reviews*, 39 (2018): 37–51, https://doi.org/10.1016/j.smrv.2017.07.001
53 Foster, *"Sleep, circadian rhythms, and health".*
54 Ibid.
55 Ibid.
56 Ibid.
57 Ibid.
58 Kelly Glazer Baron, Sabra Abbott, Nancy Jao, Natalie Manalo, and Rebecca Mullen, "Orthosomnia: Are some patients taking the quantified self too far?" *Journal of Clinical Sleep Medicine*, 13, no. 2 (2017): 351–354.
59 Foster, *"Sleep, circadian rhythms, and health".*
60 Lockley and Foster. *Sleep: A very short introduction.*
61 Ibid.

62 Cameron H. Good, Allison J. Brager, Vincent F. Capaldi, and Vincent Mysliwiec, "Sleep in the United States military." *Neuropsychopharmacology*, 45, no. 1 (2020): 176–191.
63 Pierrick J. Arnal, Fabien Sauvet, Damien Leger, Pascal Van Beers, Virginie Bayon, Clement Bougard, Arnaud Rabat, Guillaume Y. Millet, and Mounir Chennaoui, "Benefits of sleep extension on sustained attention and sleep pressure before and during total sleep deprivation and recovery." *Sleep*, 38, no. 12 (2015): 1935–1943.
64 Ibid.
65 Nicole Lovato and Leon Lack, "The effects of napping on cognitive functioning." *Progress in Brain Research*, 185 (2010): 155–166.
66 Maher Souabni, Omar Hammouda, Mohamed Romdhani, Khaled Trabelsi, Achraf Ammar, and Tarak Driss, "Benefits of daytime napping opportunity on physical and cognitive performances in physically active participants: A systematic review." *Sports Medicine* (2021): 1–32.
67 Ibid.
68 Frédéric Dutheil, Brice Bessonnat, Bruno Pereira, Julien S. Baker, Fares Moustafa, Maria Livia Fantini, Martial Mermillod, and Valentin Navel, "Napping and cognitive performance during night shifts: A systematic review and meta-analysis." *Sleep*, 43, no. 12 (2020): zsaa109.
69 Foster, *"Sleep, circadian rhythms, and health"*.
70 Sarah Laxhmi Chellappa, Roland Steiner, Peter Blattner, Peter Oelhafen, Thomas Götz, and Christian Cajochen, "Non-visual effects of light on melatonin, alertness, and cognitive performance: Can blue-enriched light keep us alert?" *PloS one*, 6, no. 1 (2011): e16429.
71 Ibid.

Jürgen Leon and Elena Trentini

When Others Have Your Back: Relational Leadership in Balint Style Groups to Enhance Stress Coping Strategies among Armed Forces

"What I look for in the speech is the response of the other"

Jacques LACAN

Although it is customary to consider the military institution as a system with its own right, with its own organization, some would be tempted to perceive it as impervious to the troubles of civil society. Such a perception is found in everyday language under the trivial expression 'military organization', which would allow those who use it, to view themselves as sheltered from uncertainty, failure and the vicissitudes of existence. However, it seems difficult to confer an infallible character on an organization that is composed of men and women who are fallible, as long as they are perceived as such.

The same observation can be made, no longer about the military organization, but about their leaders. They are not seen as weak and clumsy leader but as a protective and omnipotent. If perception is a selection of what is useful, then these qualities are those that are useful to the leader in maintaining what Bergson calls 'the veil of consciousness' in individuals.[1] We perceive the leader as infallible precisely because it is useful to us in fulfilling our need for security, which brings us back to Bergson's consideration that "to live consists in acting and accepting from objects only the useful impression in order to respond appropriately."[2] This conception easily embraces research in the field of psychology on the formation of stereotypes, described as "pictures in our heads" or "picture which often misleads men in their dealings with the world outside."[3] This categorization of information allows us to give order and predictability to our everyday life situations to act logically and efficiently, sparing cognitive resources. However, the problem is not so much the reliability of these stereotypes to reality but its correspondence to beliefs about reality. We then see that the figure of the omnipotent leader is not perceived as consensual as it seems. In this respect, they embody what makes the group come together, occupying in the minds of each member an authority figure providing protection and support. There is a shared experience, both emotional and behavioral, which social psychologists call "the common fate" and which is shared by the members of the group with regard to the idealized leader with whom they identify.[4] This also implies that when the leader falls, the group also gradually disintegrates[5] and such disorganization of troops after the capture of their leader can be observed on the battlefield. Nevertheless, psychodynamic considerations, such as those developed by Freud[6] or Bion[7] on the leader, lead us to consider that the individual evolves in a bio-psycho-social environment shaped by his or her previous experiences.

Thus, like fascination, there is also a feeling of rebellion against the authority of the leader and their parental position over the suborned. The latter are envisaged as children divided between ambivalent feelings that are the result of conflicts in which incompatible motives are involved. Sometimes loyal and devoted when the leader succeeds in generating protection and support, sometimes envious and disappointed when they fail. Thus, the figure of the leader intrigues as much as it fascinates by the permanent dialectic it provokes. This conflict will generate a certain amount of stress calling for regressive or dysfunctional defense mechanisms to deal with the situation. It is then up to the leader to contain this stress. More than the authoritarian role of the father, caring and emotional sensitivity of the mother would facilitate leader's ability to contain and manage emotions promoting functional coping strategies of the followers.

For a long time, military psychologists have studied the question of stress and coping with stressful situations through the prism of combat and war. Nowadays, armed forces are often first responders for a range of stressful interventions with repetitive exposure to potentially traumatic events and personal dangers. This may increase soldier's vulnerability to develop post-traumatic stress disorder (PTSD), related comorbidities and impaired general well-being.[8,9] However, military members are not only subjected to stressors in the context of combat and intervention (operational stressors). In fact, we often miss the full panoply of stressors, such as organizational stressors, comprising perceived unreasonableness demands of the organization or poor relationships with superordinate military leaders.[10]

The Nickname 'the Great Mute' refers to the shrinking space for speech within the armed forces.[11] Such a communication model cannot be functional, as it inhibits the relationships between the members of the organization and at the same time reinforces the occurrence of organizational stressors. These organizational stressors will be developed in this chapter and will also address a profound reflection on leadership and the tools available for its effective and innovative application to the relational field commonly found in the military.

From this point of view, a conceptual approach will be established to account for the relational dynamics at work in these stress reactions. This will enable us to put these relational processes into perspective, considering the role of the leader and how the leader acts as a container for anxiety and as a catalyst for subordinates to employ psychological resources. Optimization of supportive relationships between leader and follower will be developed accordingly. Through Balint Style Groups (BSG), we will attempt to position ourselves at the forefront of these interactions to provide a safe and tolerant space for both leaders and followers, and to promote coping strategies in the military.

Finally, we will share ideas on operationalization of BSG within the armed forces. In this respect, we will propose methods for evaluating their effectiveness and we assume this will allow the program to become established as a long-term initiative for the military. Our ambition is to prevent the stress encountered by leaders and followers and, at the same time, provide participants with tools that will allow them to enhance their relational skills with their peers in a sustainable manner.

The Intrinsic Specificity of Stress in the Armed Forces

The army is a village, an institution within an institution that has demonstrated a singular autonomy compared to another public corporation. Its commitments to a multitude of operations have made its actions particularly multifaceted. Whether it is a question of local or foreign security, logistics, human resources management, training or medical support, armed forces provide reliable and quality assistance. Such a plurality of tasks contributes to the complexity of the system and can undermine its effectiveness. This calls for flexible responses from superiors in the achievement of objectives, as the latter must consider the uncertainties of reality. Minimizing this uncertainty (i.e., the difference between the real situation and the idea of it), is ensured by coordination that allows the headquarters to ensure optimal control over different situation.[12] While flexibility allows for the increasing diversity of operational requirements within a changing strategic context, coordination ensures operational effectiveness in accordance with the actors engaged in the missions. Such requirements seriously challenge relational exchanges from decision-makers to executors. For this reason, dyadic as well as collective interactions suggest that the apparent and/or tacit difficulties of the parties involved should be considered. Being aware of the otherness of the group as well as of the individuals encourages to bring the standardization of relationships into perspective. On one hand, standardization is necessary in order to deal with difficult situations (e.g., action team phases). On the other hand, standardization will hamper the development of innovation as well as the quality of the relations between leaders and followers in situation of debriefing (e.g., transition team phases). Considering that action phases closely match with operational stressors and that transition phases match with organizational stressors, the effect of the relationship's standardization will differ. According to Nilsson,[13] favorable aspects of formal social support refer to the existence of organizational routines for communication in terms of peer support, discussion forums, debriefings, and after-action reviews. Favorable informal support includes a supportive climate promoting discussion with colleagues and to sometimes outside of the workplace. Inappropriate standardization of relations in such contexts is enough to alter the individual (Insight), interpersonal (proximity, familiarity, satisfaction) and organizational (coordination) spheres.[14] In this respect, individuals who embrace military environments metaphorically must walk on a balance beam with two sides. On one side, there are particularly stressful operations leading to homogenous practices. On the other side, there is strong institutional autonomy necessitating to be combined with relational skills, to achieve tasks in accordance with other's needs and desires. This balance exercise in which army personnel must engage has the particularity of being repetitive, specifically with structural organizational stressors. Thus, depending on the situations, a lack of flexibility in the appropriate appraisal of social interactions may constitute the gateway to an acquired vulnerability to everyday stress experienced in the army.

The Military as a Gateway to a Particular Exposure

This first reflection on structural stress in the military environment allowed us to examine the modalities of exposure to particular stimuli in more detail. The environment in which military personnel must function is incredibly challenging because they act within a reality that is constantly evolving and unpredictable. It should be noted that these events are even more striking for the military and first responders in general.[15] Indeed, as mentioned above, military members often intervene as first responders in armed conflict, which predisposes them to greater exposure to stress. At the same time, they are also expected to represent their institution with honor, which makes the armed forces subject to a particular stress exposure based on their achieved status.

First Responder's Exposition

Those known as first responders belong to a category of occupations that are often exposed to acute stress during interventions, also known as critical events. Through their deployment in armed conflicts, anti-terrorist operations, natural disasters, transportation or securing strategic locations and other stressful experiences, first responders and soldiers are comparable.[16] Thus, they are not spared from the emotional (fear, anxiety), behavioral (substance abuse, suicide) and mental health consequences (PTSD, depression, sleep disorders) that affect this category of first responder to a greater extent than non-first responders. However, although a certain number of professions can legitimately claim to be first responders, the nature and frequency of the critical events to which they are exposed seem to differ according to the missions they are assigned to. In this respect, there is an over-exposure to potentially traumatic events among military personnel, particularly for those deployed on combat operations. A reason for this is that even though first responders encounter death, it is different from how death is encountered in armed conflicts. This has the potential to significantly alter the coping strategies employed (e.g., avoidance) as well as the available social support. These are identified as protective factors for coping with the harmful effects of acute stress and which help to preserve general well-being. Nonetheless, it is important to notice that the mere fact that there is a strong exposure to potentially traumatic events need not automatically alter coping strategies but may form a basis for a need to do so. Avoidance and other reactions leading to more or less severe traumas are not desirable for soldiers, military organizations, political leadership, general public etc.

Institutional Exposition

Military personnel deployed on the battlefield or on humanitarian operations do not have a monopoly of this environment, which is comprised of a myriad of support personnel and units. The latter also includes the garrison, which involves different military requirements and whose main occupational stressors are often neglected by researchers. Nevertheless, a growing number of studies address concerns about workplace-related issues allowing us to take into account the contextual diversity to which the military is exposed[17,18,19] and, in this sense, to consider distinctive characteristics of military occupations in terms of exposure and severity of stress. Although it is possible to observe many similar stress factors between the military and civilian institutions (sleep restriction, physical exhaustion, physical injury), some stressors are particularly specific to military members. For example, we can distinguish stress due to the unique leadership climate, the ambiguity of the culture and situation, operational restrictions and language. These factors are all associated with the culture of the environment and the nature of the work demanded by the settings. In this regard, it has been noted that, within the military, the strong hierarchical culture and organizational structure per se (i.e., without social conventions, values, tacit understanding, customs etc.) are likely to reduce expectations of autonomy and control over work. This has the effect of generating stress due to organizational tension, which seems to be much more evident among garrison personnel than among combatants. As mentioned earlier, minor daily stress can have as much or even more of a deleterious effect than major stress. To better understand this distinction with respect to the effects of operational and organizational stress, it will be interesting to put into perspective the development of the two types of stress in individuals who experience them.

Related stress

Operational

Operational stress refers to the stress of critical life and death situations and events. The proximity of the event and its often brief characteristics frequently allow the qualification of acute stress to be used. This has the effect of initiating coping strategies, where soldiers seek to adapt by countering their normal resistance while attempting a return to a manageable level of stress or exhaustion. Combat operations often involve overt stress reactions, where a number of maladaptive reactions may occur. Thus, it will be common to find behaviors of stupor, agitation, panic flight and so-called automatic action. These elements make operational stress today a widely investigated factor through the prism of PTSD and other acute stress disorders encountered in post-deployment military personnel.[20,21]

Organizational

Organizational stress includes a perception of unreasonable demands imposed by the organization or a poor relationship with management.[22] It can be found among first responders as well as in soldiers in the garrison. Consequently, it is more prevalent than operational stress, which is more specific to deployed soldiers. This leads to a certain paradox: on the one hand, it concerns more people than operational stress and, on the other hand, it is the least investigated. In fact, in the context of organizational stress, it is less a question of the event itself than of all the elements that contribute to this event. There is, therefore, a combination of factors relating to the work environment's culture, leadership, and interpersonal dynamics, which, together, help to generate recurrent stress, for which consideration by peers, and individuals may be limited. Thus, we observe stress due to the leadership climate, stress due to the ambiguity of the culture and the situation, organizational restrictions, and language, which is mainly promoted by the Institution. Furthermore, a higher perception of organizational stress among military forces under interpersonal constraints have also been identified.[23]

Experienced Stress

When the person is experiencing a potentially stressful situation, the evaluation of the circumstances can determine the consequent emotional reaction. One of the main elements that influences the final emotional outcome is the appraisal, a process that helps the individual to assess the meaning of the environment for the own well-being.[24] In other words, it is a dynamic and singular evaluation of the event, allowing individuals to give a coherent emotional response according to the situation they are experiencing.

Stress as an Emotional Reaction

Depending on how the event is appraised, there will also be different changes in the intensity of felt emotion, in the physiological responses and behavioral reactions. Take the example of a young officer who must ensure the instruction of young recruits and who believes that he has correctly prepared the recruits' exercises. Initially, his emotions about this event will be positive, as it appears that the recruits are following the instructions, so he would be calm and focused on examining his recruits. However, if after the test, when talking to the recruits, he realizes that they have failed, then his emotional state about the test would change, generating a negative interpretation of that moment. This change of interpretation would lead to a feeling of sadness, a feeling of failure and the fear of not being able to pass on his knowledge, which would have an impact on the confidence he would have in future similar situations. Indeed, an appraisal change that occurs during such an emotion process would produce a domino effect on the evaluation of successive appraisals, creating subsequent negative emotional episodes.[25] Moreover, these appraisals are subjective and influenced by cultural and individual differences, leading to different emotional outcomes.[26] In fact, if two persons are confronted with the same situation, it is probable, and expected, that they would experience that moment in a different way.

All things considered, it can be said that emotions are influenced by appraisals, which in turn are influenced by other elements, such as individual differences.[27] Appraisal concept is thus important to understand the pathway of the emotional process in specific stress situation.

Stress Appraisal

Arnold[28] was one of the first theorists that included appraisal as an essential mediator between perception of the stimulus and emotion, putting the person in an active role in the evaluation of the stimulus. Through the person's evaluation of the stimulus, the emotional process takes a direction of what the reaction will be, thus preparing a fast response. This process is named action tendency, i.e., the tendency of the individual towards a specific emotional reaction after the interpretation of the situation.

Following Arnold's theory, Lazarus[29] implemented appraisals in his theory to explain emotions towards stimuli, and he structured the process in two main appraisal phases. The primary appraisal evaluates the nature of the event the person is confronted with. It gives individuals the possibility to evaluate whether the arousal is considered as positive or negative for their well-being and their goals. If the first appraisal is negatively evaluated, then the second appraisal occurs. This involves evaluating the available resources to cope with the stressful event. Depending on the availability of resources, the result can be positive (e.g., being able to cope with the event) or negative (e.g., failure to cope with the event) and this result will influence the emotional reaction of individuals.

Within the several appraisal theories, the Component Process Model (CPM) by Scherer[30] deeply describes the emotional process by using appraisals, which permit an overall and precise assessment of the event. According to Scherer, emotion is considered as an adaptive and flexible process composed of several components that interact, and the aspects involved in the emotional process are: the cognitive component (called appraisal) which evaluates internal or external stimuli; the physiological component, which captures how the autonomic nervous system will release system hormones and modify physical functioning to prepare for a coping response; the motivational component, which reflects the planning and preparing for action (action tendencies); the expression component, which allows the facial, verbal and postural communication of emotions to others; and the subjective feeling component, which monitors the interaction between individual and internal state.[31]

The CPM is structured by four main appraisal categories (Figure 1). These criteria allow the individual to evaluate the event, in the relevance, the implication, the possible coping solutions, and the normative significance of the situation. Each of these corresponds to main questions that the individual automatically answers to correctly interpret the event, called Stimulation Evaluation Checks (SECs).

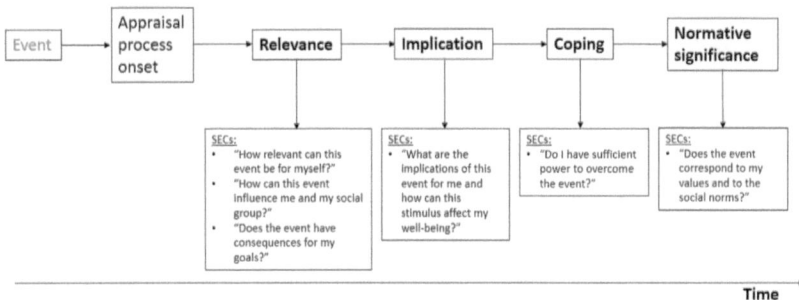

Figure 1 Graphical representation of CPM theory adapted from Sander, Grandjean and Scherer.

Sander, D., Grandjean, D., and Scherer, K. R. 2005. "A systems approach to appraisal mechanisms in emotion." *Neural Networks*, 18 (4): 317–352, https://doi.org/https://doi.org/10.1016/j.neu net.2005.03.001.

Relevance represents the degree of the importance of the stimulus for the individual. It refers to the questions: "How relevant can this event be for me?" "How can this event influence me and my social group?" and "Does the event have consequences for my goals?" Thus, using the previous example of the young officer, faced with the possible failure of his recruits, he may ask himself: "Is it really a problem if I fail to instruct them? Does this failure have consequences?"

The second step is the Implication, and it explores the possible consequences that the stimulus can have on the individual. The following corresponding question is answered: "What are the implications of this event for me and how can this stimulus affect my well-being?". Following the emotional process of the young officer, the question would be, "How much will the possible failure of my recruits stress me out? What possible consequences will this failure have in the future?"

The third goal of the evaluation process is to define the possible Coping solutions to overcome the event, through a determination of the resources. The result is an estimation of the best potential available coping strategy for the subject. One of the questions included in this category is: "Do I have sufficient power to overcome the event?" In the case of our young officer, the concrete question would be: "If I failed, do I have the resources to succeed in transmitting to them what I expect of them next time?"

The last appraisal is the Normative Significance that this event can have for the individual. It consists of the evaluation of social consequences after the reaction. The person's reaction can be strongly influenced by several ethical, moral, and social norms relating to the group to which the individual belongs.[32] It consists of the following question: "Does the event correspond to my values and to the social norms?". Returning to the young officer's situation, the question would be "Will this possible failure be socially accepted by me and by others?"

Concerning the different answers given by the person during the process, it is important to underline that the answers are not dichotomic (i.e., strictly affirmative or negative). They are based on scales or on multidimensional levels. Furthermore, the answers given by the person are subjective and depend on how the stimulus is evaluated. After all, not all young officers are the same. This implies that the intensity, and especially the type of the subsequent emotional reaction, depends on the outcomes of the different appraisals. According to Scherer,[33] the emotional process is considered as a pattern that continuously changes, producing different emotions that cannot be defined.

Development of Relational Leadership as Protective Factor

Leader's Social Support in Experienced Stress

Social support is the protective factor that moderates the harmful effects of operational and organizational stress, and it is widely viewed as an effective source of support in the literature.[34] It constitutes a function that can be exercised by both individuals and the Institution, especially within the military environment. Indeed, it has been found that social support deficiencies in military samples are associated with more negative psychological effects than in civilian samples.[35] This indicates that the armed forces are a population for which the availability and provision of support must be a central concern. Thus, the environment in which they work must allow for working conditions that are adapted to the type of their mission. Moreover, social support in relation to the environment (general embeddedness) seems to be an important protective factor in preventing the development of PTSD as well as for mental health in general.[36,37] Given this, it will be important for leaders to position themselves as mediators between the environment and their subordinates. Indeed, leaders must make the protective resources available within an appropriate work environment. Thus, it will be important for military members to be aware of the elements that facilitate specific interactions buffered by leader social support between the environment and their subordinates, which is an integral part of the leader's role and influences leadership effectiveness.

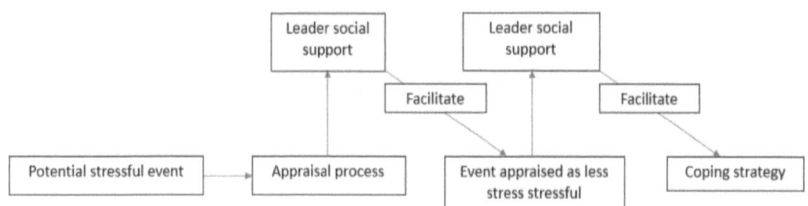

Figure 2 Adapted version of Cohen's Buffering model with leader social support.
Cohen, S. and Wills, T. A. 1985. "Stress, social support, and the buffering hypothesis." *Psychological Bulletin*, 98 (2): 310–357, https://doi.org/10.1037/0033-2909.98.2.310

Leader support has been conceptualized in accordance with Cohen's[38] buffer model (Figure 2). This model established two pathways for leader social support occurring at the time of deployment and after deployment of stressors. The first prevents the negative appraisal of stressors, while the second encourages reappraisal and facilitates adaptive responses. According to the model, an individual experiencing specific stressors is best protected by leadership support that provides stressor-specific coping resources, such as counselling or cognitive guidance. In this view, actual social exchanges, and relative support transactions (experienced or observed) help to form and maintain the perception of a supportive social environment.

In this regard, interpersonal relationships are also a strategic channel of communication to promote social support. This can easily be envisaged by developing the relationship between leader and followers, as supported by Leader-Member Exchange theory (LMX theory). The enhancement of this exchange dynamic between the parties highlights their mutual influence and facilitates reflections on their respective roles. Depending on the quality of the exchanges maintained, the coping strategies will be reinforced by facilitated evaluations of the stress situations that can bother the evaluation model that we developed above.

A bio-psycho-social understanding of such leader-follower relationships is facilitated by leadership training programs. Among the latter, BSG is a privileged way to consider these relationships thanks to the secure and tolerant space it provides. This program is an innovative and promising tool that seems to be adapted to the reinforcement of coping strategies but also relevant to the development of leadership. Thus, the environment, the relationships but also the training programs appear as effective, available, and applicable tools aiming at consolidating the highly beneficial social support in the prevention of the consecutive failures of the stressful situations. After having integrated the relational leadership theory to the leader support model and the effects on coping strategies, we will further present the practice and the benefits for the armed forces of a BSG program.

Relational Leadership, LMX Theory

The issue of leadership has always been considered fundamental to the armed forces, as evidenced by the abundance of literature in this area.[39,40] Nevertheless, most of the studies that attempt to address the concept of leader social support to military mental health have chosen to investigate the social support provided by leadership in a quantitative manner.[41] This further reveals the scarcity of qualitative approaches to the topic. For example, examining relevant support relationships remain extremely limited, even as consistent findings emerge regarding the protective role of leader support.[42,43] This is likely the consequence of an evolution over the past several decades of the role of the leader. Early conceptions were quickly popularized by a dispositional approach echoing what are still called "great-man theories."[44] From the middle of the 20th century, a behavioral approach was developed in which being a leader is not a question of characteristics, but rather a way of acting. Then, a situational approach was also considered, which explored leadership within the context of environmental factors. More recently, the relational approach has focused on the interactions at work between the members and the leader, both at the dyadic and collective levels. Leader-follower interactions can therefore be described in terms of a process (LMX) producing certain effects that are perceptible as the relationship matures (role taking- role making-routinization). This is in the same trend as transformational leadership, which Guimond[45] wonderfully described as "the set of activities and especially communications by which an individual influences the behavior of the members of a group in the direction of the voluntary realization of certain common objectives."[46] In other words, no one comes leader, but we can become a leader if we exercise an influence over other. If we had to summarize, we could say that in reality it teaches us that we are not born leaders, but that we can become a leader if we exercise an influence over others. However, the outcome of this process requires taking into account the dynamics that make it up. In this respect, LMX theory assumes that leaders do not treat all members in the same way and that this will depend on the quality of the relationships maintained with them. This will then push the leader to categorize followers as in-group or out-group which will then describe two different realities where for some the quality of LMX will be high (in-group) and for others it will be low (out-group). This will then have consequences in terms of work attitude, communication, and performance for the group members. A high LMX quality has been associated with greater trust in the group, more interaction, support, and reward from the leader. Furthermore, this quality of communication allows for stronger social bonds, increased problem solving and regulation of emotions.[47] On the other hand, individuals belonging to the out-group will show an impoverishment of communication, will be more distant from the leader, and will be less ardently involved in their mission according to a *Come-Work-Leave* logic.

Applied to the armed forces, this makes sense insofar as the military environment represents one of the paragons of leadership expression. It is therefore obvious in this respect that the management of human conflicts, personal difficulties as well as the cohabitation of teams are the business of military leaders. By allowing their subordinates to deal with these issues through relational leadership, it allows them to put the human being back at the center, as suggested by General de Villiers.[48] By doing so, the subordinates manage to extricate themselves from a low, infantilizing position, and to raise themselves to the state of an adult, which gives them the necessary space to take on responsibilities that will continue to facilitate the work of the leaders. The concrete effects of this act of empowerment in term of satisfaction, performance, role- and conflict clarity[49] among followers are perfectly visible in the LMX theory that encourages leaders to be attentive to the quality of their relationships with all their subordinates and not just the closest ones. If they feel supported, they will be able to deploy the resources that lead them to greater autonomy.

Coping Strategies' Reinforcement

Coping refers to the intentional efforts we engage in to minimize the physical, psychological, or social harm of situation. Literature tends to distinguish between two main coping strategies: Problem-focused (based on one's ability to manage the environmental event) and Emotion-focused (focus on changing the emotions caused by a stressful event).[50] People who cannot manage their stress level at work may harm themselves or others. For this reason, strengthening coping strategies is a good approach to relieve leader's and follower's stress effectively and ensure work efficiency as well as constructive relationships.[51] This approach, when integrated with Cohen's buffer model[52] and completed by a functional evaluation of the situation, offers space and great autonomy to the individuals concerned.

Problem-focused coping included direct action, instrumental social support seeking and preparatory action. They are mainly recommended to cope with organizational stress[53,54] and allowed individual to change the situation itself. Considering that, reinforcing these strategies therefore highly relevant within the garrison, previously described as more vulnerable to organizational stress.

Direct action refers to leader's taking direct action to remove the stressor in order to resolve the problem.[55] By doing so, direct action may moderate stressful situation and would lead to enhance general well-being.[56,57]

Preparatory action is the ability to choose among different ways of proceeding with the evaluation of choices.[58] This is facilitated by a great communication between team members by taking into account the opinions of other members (in-group and out-group). As a result, it would lead to a greater team spirit as well as a greater job performance.[59]

Instrumental social support seeking is the ability to acquire information about the stressful event from different people.[60] This is particularly important because it is a proactive coping strategy that allows people to anticipate stressful situations to minimize difficulties before they occur. Individuals engaged in such coping strategies tend to perceive stressful events as growth opportunities and tends also to communicate more frequently, reinforcing workplace's relationships.[61]

Emotion-focused coping aimed to adjust emotional distress and maintaining a moderate level of arousal in order to not be overwhelmed by stress by emotion.[62]

Among the most investigated emotion-focused coping strategies is emotional regulation. It consists of a direct behavior that a person will initiate in order to trigger a succession of physiological, biological and psychological processes. For example, muscle relaxation through deep breath control during armful situation has been found in several studies to be an effective strategy to reduce stress.[63,64]

Cognitive restructuring is also part of the palette of reinforced emotion-focused coping strategies;[65] we speak of cognitive restructuring when the individual can think differently (i.e., positively) about the emotions triggered in response to a stressful situation.[66] By attempting to change one's view of a stressful situation with positive thinking, individual will reduce anxiety and increased self-efficacity.[67]

Seeking emotional support involves talking with someone else (leader, BSG, follower) in order to get moral support, sympathy or understanding.[68]

Considering that these two-way coping strategies studies found that problem-focused behaviors are often the dominant coping response when individual perceives a moderate level of stress.[69] Problem-focused strategies seems to be predominant in organizational stress situation while emotion focused strategies predominated in highly stress situation, leading to insufficient attention to problem solving mechanisms.[70] However, as noted by Cohen,[71] an effective coping strategy does not only resolve the problem but also facilitate positives consequences, reduce the impact of stress, and moderate its negative consequences. For these reasons, coping usually incorporated emotion-focused and problem focused strategies.

This two-way coping strategy (emotion and problem focused) helps leaders and followers to cope with situations in a more flexible way, which is a particularly helpful quality in the military environment where flexibility is a structural component according to the stressors. Nevertheless, many leaders do not exhibit the propensity to apply such combination of strategies which impairs effective leadership and their relations with followers. Thus, this raises a question concerning the effective transmission of such knowledge that contributes to strengthening a relationship that promotes the quality of exchanges through leadership training. We have begun to respond to this by setting up a specific training program called BSG which will be the transmission belt of the relational leadership that we aim to develop within the armed forces.

BSG Address for Leaders of Armed Forces

The effective training programs depends in part on organizational conditions that facilitates or inhibit learning of leadership skills and the application of this learning by leaders.[72] A comprehensive review of the literature found that BSG participants (>1.5 years) had significantly higher scores on self-reported control, satisfaction, job quality, and health.[73] However, none of them have explored the relevance of these supportive relationships in the military domain. This article aims to be the first to provide a review of the beneficial outcomes of these BSGs for coping with high-stress situations, which are extremely prevalent among first responders in the military.

Recommendations will be given for operationalizing and evaluating BSG for military members. Finally, critical leadership issues will be discussed, including leaders' experiences with challenges in their groups, as well as important skills attributed to BSG leadership.

BSG as a Tool to Cope with Stress

BSG method consists of exploring the difficulties that general practitioners (GPs) encounter in their interactions with their patients through discussions in regular group sessions composed of other professionals (usually between 6 and 12 people) and one or more facilitators (also called Balint leaders). The goal is to let the group members express themselves on a specific case and thus share the emotions and thoughts that the case evokes for them, which favors a reflective approach to their practice. Balint leaders aim to promote a safe and tolerant environment for participants to express themselves without fear of value judgments that would be detrimental to free expression. The frequency of these groups is typically established over a long term, and it is common for these group sessions to extend over several years. Studies tend to show a significant benefit from participating as a BSG member after 1.5 years within this sort of group setting.

These groups were initially developed for general practitioners but were quickly expanded to include different groups of professionals with leadership responsibilities. In this respect, the BSGs revealed a capacity to provoke a common and collegial reflection in the expression of the roles that the leaders' play in difficult situations as well as helping them to analyze their interaction and their relationship with their followers. The method is then seen as a dynamic process that puts professional factors into perspective by establishing organizational and personal goals through the construction of a space conducive to work relationships.[74] The situations initially evaluated as stressful by the participants are thus likely to be re-evaluated according to the discussions generated by the group members. Thus, BSGs allow for a direct influence on the different categories of SECs[75] and thus play a strategic role in the flexibility of the emotional process. It is by letting others provide support in the appreciation of these stressful situations that this method represents a useful means of strengthening coping strategies. However, BSGs do not only strengthen coping strategies from a leadership or organizational point of view. In fact, it counts as a coping strategy and might be sometimes the only coping strategy that is necessary. Studies show that Balint participants felt more in control of the work situation and felt more able to cope with feelings of helplessness[76] compared to non-Balint participants. Important to note is that this difference increased with longer time in Balint groups. These two elements underline the importance of BSG continuing for at least 1.5 years, infused by confidentiality, allowing the establishment of an atmosphere of trust and loyalty. Trust and loyalty are particularly cherished virtues within the armed forces but are often limited to smaller groups and contribute to enhance development of strong group identities.[77] By supporting reflexive thinking BSG aims to develop leaders' awareness on such aspects of trust and loyalty not just in the in-group but also in the out-group. As a result, a deeper understanding of human interactions and leaders personal influence on the in- and out-groups may also help to strengthen the spirit. In addition to relational empowerment, BSGs aim to stress problem- as well as emotion focused strategies thought the interaction between members.

Operationalization of BSG in the Armed Forces

It is not easy to train leaders who can both contain collective fears and meet the needs of their subordinates by providing them with the autonomy to be empowered. Nonetheless, given the closeness and comradeship among members of the military, it is reasonable to assume that this environment provides particularly fertile ground for the creation of reflective leaders. On other hand, closeness and comradeship can give rise to group think with hinders reflective leadership.[78] In this regard, it is important to consider several elements to put in place when operationalizing a BSG program

- Periodicity: agree on weekly or monthly sessions (consult with group members on frequency).
- Location: same for each session and away from prying eyes
- Duration: between 1.5 and 2 hours per session
- Group size: between 6 and 12 participants
- Composition of the group:
    o Participants: Individuals with similarities in tasks performed.
    o Balint facilitator: Individuals familiar with the concepts of group psychodynamics or military personnel who have already successfully experienced a BSG with their peers for a sufficient period of time (>1.5 years).
- Program duration: Minimum of 1.5 years
- Group Flexibility: Semi-open group (open to new entries if maximum quotas are not met and all group members agree).
- Process: Welcome the participants – Brief reminder of the previous session – Spontaneous presentation of a situation/conflict – Discussion of the emotions and thoughts evoked with free and kind words – Synthesis of the elements communicated around the shared situation – Thanks and closing of the session.

The evaluation of the situation/conflict stressed by leaders in a BSG can be done at several level by the participants. Thus, some strategies of conflict resolution apply adequately at this a certain level but are not very effective for the others. If one neglects to consider an important level of the conflict, one runs the risk of striving to apply sterile strategies that are ineffective.

The personal level explained the conflict on the basis of the personal characteristics of one or other of the protagonists (aggressiveness, incompetence, rudeness, etc…)

Relational level explained conflict from the nature of a relationship (disturbed communication, dominance ...). The methods of resolution here can call upon the techniques of relational therapy (Systemic, transactional)

Group level explained the conflict by an atmosphere specific to the group dynamic (rivalry, distribution of power, pressure, role of the leader).

Organizational level explained conflict from an organizational structure: schedules distribution of premises, environmental context, rules. Here it is a question of arriving at a reflection on the organizational functioning.

Finally, institutional level explained the conflict from law, norms, values, ideological or cultural models that underlie the behaviors (competitiveness, paternalism, sexism...). The conflict may arise, for example, from an incompatibility between the values displayed and the method of work. The resolution will have to focus on transposing the ideological conflict into a methodological conflict.

BSG and Consequences for Leadership Development

By developing such a program for leaders, several positive effects are expected for them. One of the primary consequences is to provide them the opportunity to reflect on their work and the relationships they have within this environment. It also provides them with an outlet for the anxieties and frustrations associated with their work environment. From a relational standpoint, it offers the possibility of stimulating the interest that leaders may have in individuals with whom they have had difficult interactions. It also offers leaders a new way of looking at day-to-day management. The group has their back by offering support that allows leaders to improve communication. Thus, the consequences of such program are to facilitate the way in which leaders interact and thus result in a better understanding of the Other and help develop emotional adaptability[79] in human relationships. Among other key skills are the knowledge of facilitating techniques (bringing back forgotten interventions, making links between contributions), clarifying what is happening in the group, exploring the relationships between group members, and the ability to reflect the emotions and feelings of the members.

Conclusion

We began this chapter with an explanation of what makes the armed forces specific, mentioning their flexibility and coordination in the face of the plurality of missions encountered. In this respect, the military environment subjects its members to special conditions, whether on intervention operations or, more generally, in the garrison, where the hierarchical structure and strict rules constitute a distinct institutional framework. This generates different types of stress categorized as operational, mostly consisting of acute phases, and organizational, characterized by a multiplicity in the occurrence of minor stress phases. To understand the effects and the margin of action available to individuals, we have highlighted stress as an emotional reaction and then explained its dynamics in terms of evaluation for the individual. A brief history of the concepts around the evaluation led us to include the CPM,[80] whose major advantage is that the configuration of SECs allows one to predict the emotional reaction. We then addressed the question of the leader's support in the face of stress based on Cohen's buffer model,[81] which gives us insight into the leader's actionable moments during the stages of the stress reaction. Relational leadership through LMX theory presented an effective and relevant leadership style in the military environment given the tendency to develop acquired vulnerability to organizational stressors and interpersonal relationship deficit often reported in the literature. Thus, the benefits of such a type of leadership adopted by the armed forces permits a diversity the types of coping strategies. To do this, we conceptualized relational leadership through BSGs, which, in addition to being innovative within the armed forces, also have the advantage of offering a particularly fertile ground for the development of such leadership. Ways of operationalizing the BSG program have been proposed along the lines just mentioned, including keys parameters in term of procedure and duration. Finally, more specific considerations regarding the skills developed by the leaders were addressed to account for the potential of such a program within the armed forces.

The path we have initiated around the issue of stress in the armed forces has enabled us to develop a training program for leaders. We can conclude that BSG is a real innovation and that it deserves to be tested. It provides both problem-focused and emotionally reflective strategies and encourages a better understanding of the relationships between people. Training today and tomorrow's leaders is a beneficial investment for organizations and the individuals within them. Whether it is the leaders and their relational and reflective potential or the followers and their growing autonomy, the BSG promotes an alternative consideration of stressful situations for everyone.

Notes

1 H. Bergson, *Le rire*. Flammarion, 2013.
2 Ibid, p. 81.
3 W. Lippmann, The world outside and the pictures in ours heads. In *Public opinion*, pp. 3–32, p. 21 (1922). New York, NY, US: MacMillan Co.

4 S. Fiske, *Psychologie sociale*, p. 586 (2008). De Boeck Supérieur.
5 S.A. Wheelan, M. Åkerlund, and C. Jacobsson, *Creating effective teams: A guide for members and leaders* (2020). Sage Publications.
6 S. Freud, *Totem and taboo*, Vol. 13, *Standard edition*, (1912). London.
7 W.R. Bion, *Experiences in groups and other papers* (1961). London: Tavistock Publications.
8 W. Berger, E.S.F. Coutinho, I. Figueira, C. Marques-Portella, M.P. Luz, T.C. Neylan, . . . M.V. Mendlowicz, "Rescuers at risk: A systematic review and meta-regression analysis of the worldwide current prevalence and correlates of PTSD in rescue workers." *Social Psychiatry and Psychiatric Epidemiology*, 47, no. 6 (2012): 1001–1011, https://doi.org/10.1007/s00127-011-0408-2
9 S. Jones, "Describing the mental health profile of first responders: A systematic review." *Journal of the American Psychiatric Nurses Association*, 23, no. 3 (2017): 200–214, https://doi.org/10.1177/1078390317695266
10 Jean-Baptiste Antoine Marcellin De Marbot, *Mémoires du général baron de Marbot* (1891). Mercure de France.
11 M.O. Heisler, "The French army in politics, 1945–1962." Review of the French army in politics, 1945–1962, John Steward Ambler. *World Affairs*, 129, no. 4 (1967): 268–270, http://www.jstor.org/stable/20670853
12 Dan Horowitz, "Flexible responsiveness and military strategy: The case of the Israeli Army." *Policy Sciences*, 1, no. 2 (1970): 191–205, http://www.jstor.org/stable/4531385
13 A. Nilsson and M. Engström, "E-assessment and an e-training program among elderly care staff lacking formal competence: Results of a mixed-methods intervention study." *BMC Health Services Research*, 15, no. 1 (2015): 189, https://doi.org/10.1186/s12913-015-0843-y
14 R.E. Kraut, R.S. Fish, R.W. Root, and B. L. Chalfonte, "Informal communication in organizations: Form, function, and technology." In S. Oskamp and S. Spacapan (Eds.), *Human reactions to technology: The claremont symposium on applied social psychology* (1990). Beverly Hills, CA: Sage.
15 A.Walker, A. McKune, S. Ferguson, D.B. Pyne, and B. Rattray, "Chronic occupational exposures can influence the rate of PTSD and depressive disorders in first responders and military personnel." *Extreme Physiology & Medicine*, 5, no. 1 (2016): 8, https://doi.org/10.1186/s13728-016-0049-x
16 D.M. Benedek, C. Fullerton, and R.J. Ursano, "First responders: Mental health consequences of natural and human-made disasters for public health and public safety workers." *Annual Review of Public Health*, 28 (2007): 55–68, https://doi.org/10.1146/annurev.publhealth.28.021406.144037
17 S.C. Wattie and R.S. Bridger, "Work-related stress indicator surveys in UK Ministry of Defence." *Journal of the Royal Army Medical Corps*, 165, no. 2 (2019): 128–132, https://doi.org/10.1136/jramc-2018-001042

18 G. Larsson, A.K. Berglund, and A. Ohlsson, "Daily hassles, their antecedents and outcomes among professional first responders: A systematic literature review." *Scandinavian Journal of Psychology*, 57, no. 4 (2016): 359–367, https://doi.org/https://doi.org/10.1111/sjop.12303
19 A.Walker, A. McKune, S. Ferguson, D.B. Pyne, and B. Rattray, "Chronic occupational exposures can influence the rate of PTSD and depressive disorders in first responders and military personnel." *Extreme Physiology & Medicine*, 5, no. 1 (2016): 8, https://doi.org/10.1186/s13728-016-0049-x
20 H. Wang, H. Jin, S. E Nunnink, W.Guo, J.Sun, J Shi, . . . D.G. Baker, "Identification of post traumatic stress disorder and risk factors in military first responders 6months after Wen Chuan earthquake in China." *Journal of Affective Disorders*, 130, no. 1 (2011): 213–219, https://doi.org/https://doi.org/10.1016/j.jad.2010.09.026
21 D.M. Benedek, C. Fullerton, and R.J. Ursano, "First responders: mental health consequences of natural and human-made disasters for public health and public safety workers." *Annual Review of Public Health*, 28 (2007): 55–68, https://doi.org/10.1146/annurev.publhealth.28.021406.144037
22 Hsieh, Chi-Ming and Tsai, Bi-Kun., "Effects of social support on the stress-health relationship: Gender comparison among military personnel." *International Journal of Environmental Research and Public Health*, 16, no. 8 (2019): 1317, https://doi.org/10.3390/ijerph16081317
23 Hsieh, Chi-Ming and Tsai, Bi-Kun, "Effects of social support on the stress-health relationship: Gender comparison among military personnel." *International Journal of Environmental Research and Public Health*, 16, no. 8 (2019): 1317, https://doi.org/10.3390/ijerph16081317
24 A. Moors, P.C. Ellsworth, K.R. Scherer, and N.H. Frijda, "Appraisal theories of emotion: State of the art and future development." *Emotion Review*, 5, no. 2 (2013): 119–124, https://doi.org/10.1177/1754073912468165
25 K.R. Scherer, and A. Moors, "The emotion process: Event appraisal and component differentiation." *Annu Rev Psychol*, 70 (2019): 719–745, https://doi.org/10.1146/annurev-psych-122216-011854
26 K.R. Scherer, "The dynamic architecture of emotion: Evidence for the component process model." *Cognition and Emotion*, 23, no. 7 (2009): 1307–1351, https://doi.org/10.1080/02699930902928969
27 A. Moors, P.C. Ellsworth, K.R. Scherer, and N.H. Frijda, "Appraisal theories of emotion: State of the art and future development." *Emotion Review*, 5, no. 2 (2013): 119–124, https://doi.org/10.1177/1754073912468165
28 M.B. Arnold, *Emotion and personality. Emotion and personality* (1960). New York, NY, US: Columbia University Press.
29 R.S. Lazarus, "Progress on a cognitive-motivational-relational theory of emotion." *American Psychologist*, 46, no. 8 (1991): 819–834, https://doi.org/10.1037/0003-066X.46.8.819

30 K.R. Scherer, "On the nature and function of emotion: A component process approach." In K.R. Ekman Scherer (Ed.), *Approaches to emotion*, pp. 293–317 (1984). Hillsdale: Erlbaum.
31 K.R. Scherer, "What are emotions? And how can they be measured?" *Social Science Information*, 44, no. 4 (2005): 695–729, https://doi.org/10.1177/0539018405058216
32 R. Greco, Component process model: Una teoria dinamica dell'emozione. *Cognitivismo Clinico*, 7, no. 2 (2010): 124–141, https://www.apc.it/wp-content/uploads/2013/06/04-greco.pdf
33 K.R. Scherer, "On the nature and function of emotion: A component process approach." In K.R. Ekman Scherer (Ed.), *Approaches to emotion*, pp. 293–317 (1984). Hillsdale: Erlbaum.
34 R. Geuzinge, M. Visse, J. Duyndam, and E. Vermetten, "Social embeddedness of firefighters, paramedics, specialized nurses, police officers, and military personnel: Systematic review in relation to the risk of traumatization." *Frontiers in Psychiatry*, 11, no. 1450 (2020), https://doi.org/10.3389/fpsyt.2020.496663
35 C.R. Brewin, B. Andrews, and J.D. Valentine, "Meta-analysis of risk factors for posttraumatic stress disorder in trauma-exposed adults." *Journal of Consulting and Clinical Psychology*, 68, no. 5 (2000): 748–766, https://doi.org/10.1037//0022-006x.68.5.748
36 A. Ahronson, and J.E. Cameron, "The nature and consequences of group cohesion in a military sample." *Military Psychology*, 19, no. 1 (2007): 9–25, https://doi.org/10.1080/08995600701323277
37 J. Breslau, C.M. Setodji, and C.A. Vaughan, "Is cohesion within military units associated with post-deployment behavioral and mental health outcomes?" *Journal of Affective Disorders*, 198 (2016): 102–107, https://doi.org/https://doi.org/10.1016/j.jad.2016.03.053
38 S. Cohen, and T.A. Wills, "Stress, social support, and the buffering hypothesis." *Psychological Bulletin*, 98, no. 2 (1985): 310–357, https://doi.org/10.1037/0033-2909.98.2.310
39 G.L. Siebold, "The relation between soldier motivation, leadership, and small unit performance." *Motivation theory and research* (1994).
40 J.M. Savell, R.C. Teague, and T. R. Tremble Jr, "Job Involvement Contagion Between Army Squad Leaders and Their Squad Members." *Military Psychology*, 7, no. 3 (1995): 193–206, https://doi.org/10.1207/s15327876mp0703_3
41 R. Geuzinge, M. Visse, J. Duyndam, and E. Vermetten, "Social embeddedness of firefighters, paramedics, specialized Nurses, Police Officers, and Military Personnel: systematic review in relation to the risk of traumatization." *Frontiers in Psychiatry*, 11, no. 1450 (2020), https://doi.org/10.3389/fpsyt.2020.496663
42 K. Brailey, J.J. Vasterling, S.P. Proctor, J.I. Constans, and M.J. Friedman, "PTSD symptoms, life events, and unit cohesion in U.S. soldiers: Baseline findings

from the neurocognition deployment health study." *Journal of Traumatic Stress*, 20, no. 4 (2007): 495–503, https://doi.org/10.1002/jts.20234

43 J. Du Preez, J. Sundin, S. Wessely, and N.T. Fear, "Unit cohesion and mental health in the UK armed forces." *Occupational Medicine*, 62, no. 1 (2011): 47–53, https://doi.org/10.1093/occmed/kqr151

44 P.G. Northouse, *Leadership: Theory and practice* (1999). E-Content Generic Vendor.

45 S. Guimond, "Les groupes sociaux in Vallerand." In Gaëtan Morin (Ed.), *Les fondements de la psychologie sociale* (1994). Montréal.

46 Ibid., p. 79.

47 S. Guimond, "Les groupes sociaux in Vallerand." In Gaëtan Morin (Ed.), *Les fondements de la psychologie sociale* (1994). Montréal.

48 P. De Villiers, *Qu'est-ce qu'un chef?: Pluriel* (2019).

49 C.R. Gerstner and D.V. Day, "Meta-analytic review of leader–member exchange theory: Correlates and construct issues." *Journal of Applied Psychology*, 82, no. 6 (1997): 827–844, https://doi.org/10.1037/0021-9010.82.6.827

50 C.S. Carver, M.F. Scheier, and J.K. Weintraub, "Assessing coping strategies: A theoretically based approach." *Journal of Personality and Social Psychology*, 56, no. 2 (1989): 267–283, https://doi.org/10.1037/0022-3514.56.2.267

51 Mei-Yung Leung, A.M.M. Liu, and M. Mei-ki Wong, "Impact of stress-coping behaviour on estimation performance." *Construction Management and Economics*, 24, no. 1 (2006): 55–67, https://doi.org/10.1080/01446190500228381

52 S. Cohen and T.A. Wills, "Stress, social support, and the buffering hypothesis." *Psychological Bulletin*, 98, no, 2 (1985): 310–357, https://doi.org/10.1037/0033-2909.98.2.310

53 J.N. Tillmann and M.T. Beard, "Manager's healthy lifestyles, coping strategies, job stressors and performance: An occupational stress model." *Journal of Theory Construction & Testing*, 5, no. 1 (2001): 7, https://www.proquest.com/openview/060bce10ce9175946bf16354ae29d128/1?cbl=11511&pq-origsite=gscholar

54 Mei-Yung Leung, A.M.M. Liu, and M. Mei-ki Wong, "Impact of stress-coping behaviour on estimation performance." *Construction Management and Economics*, 24, no. 1 (2006): 55–67, https://doi.org/10.1080/01446190500228381

55 T.J. Newton and A. Keenan, "Coping with work-related stress." *Human Relations*, 38, no. 2 (1985): 107–126, https://doi.org/10.1177/001872678503800202

56 T.J. Newton and A. Keenan, "Coping with work-related stress." *Human Relations*, 38, no. 2 (1985): 107–126, https://doi.org/10.1177/001872678503800202

57 L. Fortes-Ferreira, J.M. Peiró, M.G. González-Morales, and I. Martín, "Work-related stress and well-being: The roles of direct action coping and palliative coping." *Scandinavian Journal of Psychology*, 47, no. 4 (2006): 293–302, https://doi.org/https://doi.org/10.1111/j.1467-9450.2006.00519.x

58 C.S. Carver, M.F. Scheier, and J.K. Weintraub, "Assessing coping strategies: A theoretically based approach." *Journal of Personality and Social Psychology*, 56, no. 2 (1989): 267–283, https://doi.org/10.1037/0022-3514.56.2.267

59 Mei-Yung Leung, A.M.M. Liu, and M. Mei-ki Wong, "Impact of stress-coping behaviour on estimation performance." *Construction Management and Economics*, 24, no. 1 (2006): 55–67, https://doi.org/10.1080/01446190500228381

60 F.M. Sterle, T. Vervoort, and L.L. Verhofstadt, "Social support, adjustment, and psychological distress of help-seeking expatriates." *Psychologica Belgica*, 58, no. 1 (2018): 297–317, https://doi.org/10.5334/pb.464

61 E.R. Greenglass, "Proactive coping and quality of life management." In *Beyond coping: Meeting goals, visions, and challenges*, pp. 37–62 (2002). New York, NY, US: Oxford University Press.

62 R. Djebarni, "The impact of stress in site management effectiveness." *Construction Management and Economics*, 14, no. 4 (1996): 281–293, https://doi.org/10.1080/014461996373368

63 R. McCraty and M.A. Zayas, "Cardiac coherence, self-regulation, autonomic stability, and psychosocial well-being." *Frontiers in Psychology*, 5, no. 1090 (2014), https://doi.org/10.3389/fpsyg.2014.01090

64 J.P. Ginsberg, M.E. Berry, and D. A Powell, "Cardiac coherence and posttraumatic stress disorder in combat veterans." *Alternative Therapies in Health and Medicine*, 16, no. 4 (2010): 52–60, https://pubmed.ncbi.nlm.nih.gov/20653296/.

65 G.M. De Boo and J.M. Wicherts, "Assessing cognitive and behavioral coping strategies in children." *Cognitive Therapy and Research*, 33, no. 1 (2007): 1, https://doi.org/10.1007/s10608-007-9135-0

66 E.A., Skinner, K. Edge, J. Altman, and H. Sherwood, "Searching for the structure of coping: A review and critique of category systems for classifying ways of coping." *Psychological Bulletin*, 129, no. 2 (2003): 216–269, https://doi.org/10.1037/0033-2909.129.2.216

67 C.J. Haney, "Stress-management interventions for female athletes: Relaxation and cognitive restructuring." *International Journal of Sport Psychology*, 35, no. 2 (2004): 109–118, https://psycnet.apa.org/record/2004-17520-002.

68 T.J. Newton and A. Keenan, "Coping with work-related stress." *Human Relations*, 38, no. 2 (1985): 107–126, https://doi.org/10.1177/001872678503800202

69 C.R. Anderson, "Coping behaviors as intervening mechanisms in the inverted-U stress-performance relationship." *Journal of Applied Psychology*, 61, no. 1 (1976): 30–34, https://doi.org/10.1037/0021-9010.61.1.30
70 A.P. Wolfgang, "Job stress, coping, and dissatisfaction in the health professions: A comparison of nurses and pharmacists." *Journal of Social Behavior and Personality*, 6, no. 7 (1991): 213, https://www.proquest.com/scholarly-journals/jobb-stress-coping-dissatisfaction-health/docview/1292262854/se-2?accountid=12006
71 S. Cohen, "Aftereffects of stress on human performance and social behavior: A review of research and theory." *Psychological Bulletin*, 88, no. 1 (1980): 82–108, https://doi.org/10.1037/0033-2909.88.1.82
72 G.A. Yukl, *Leadership in organizations*, 6th ed. (2005) Upper Saddle River, NJ: Prentice Hall.
73 K. Van Roy, S. Vanheule, and R. Inslegers, "Research on Balint groups: A literature review." *Patient Education and Counseling*, 98, no. 6 (2015): 685–694, https://doi.org/10.1016/j.pec.2015.01.014.
74 A. Mastrangelo, E.R. Eddy, and S.J. Lorenzet, "The importance of personal and professional leadership." *Leadership & Organization Development Journal*, 25, no. 5 (2004): 435–451, https://doi.org/10.1108/01437730410544755.
75 K.R. Scherer, "The dynamic architecture of emotion: Evidence for the component process model." *Cognition and Emotion*, 23, no. 7 (2009): 1307–1351, https://doi.org/10.1080/02699930902928969.
76 Dorte Kjeldmand and Inger Holmström, "Balint groups as a means to increase job satisfaction and prevent burnout among general practitioners." *The Annals of Family Medicine*, 6, no. 2 (2008): 138–145, https://doi.org/10.1370/afm.813.
77 J. Connor, D.J. Andrews, K. Noack-Lundberg, and B. Wadham, "Military loyalty as a moral emotion." *Armed Forces & Society*, 47, no. 3 (2021): 530–550, https://doi.org/10.1177/0095327x19880248.
78 J. Connor, D.J. Andrews, K. Noack-Lundberg, and B. Wadham, "Military loyalty as a moral emotion." *Armed Forces & Society*, 47, no. 3 (2021): 530–550, https://doi.org/10.1177/0095327x19880248.
79 S. David, *Emotional agility* (2013). New York: Avery.
80 K.R. Scherer, "The dynamic architecture of emotion: Evidence for the component process model." *Cognition and Emotion*, 23, no. 7 (2009): 1307–1351, https://doi.org/10.1080/02699930902928969.
81 S. Cohen and T.A. Wills, "Stress, social support, and the buffering hypothesis." *Psychological Bulletin*, 98, no. 2 (1985): 310–357, https://doi.org/10.1037/0033-2909.98.2.310

Verma Swati, Updesh Kumar and Dakshi Walia

Post Traumatic Growth in Military Personnel

> *Between stimulus and response, there is a space. In that space is our power to choose our response. In our response lies our growth and freedom. – Victor Frankl[1]*

The goal of every military in the world is to strengthen their forces physically and mentally, so they design and offer various training programs. Regardless of how well prepared soldiers are, the nature of the job is such that they will encounter traumatic situations. A retrospective reflection on the long-term consequences of any stressful or traumatic event helps us understand both the good and bad aspects of that event. Negative events sometimes lead to positive change, like recognizing one's strength, improved relationships with spouse, children or colleagues, an ardent desire to explore new avenues, spiritual growth and a greater appreciation of life. The year 2020 witnessed a pandemic, which is still lingering on at the time of writing this chapter. Many people, after experiencing a stressful or traumatic situation, have attempted to find some positive out of such crisis, thereby exhibiting the endowing human spirit to survive and thrive.

Stress resulting from war or being in captivity as prisoner of war is huge and multi-layered. Much of the stress soldiers sustain is interpersonal in nature, caused by psychological abuse, humiliation and punitive torment.[2] As a result, most of the studies pertaining to war induced trauma focus on its negative consequences. Many researchers have investigated the pernicious after-effects of war and captivity in terms of post-traumatic stress disorder (PTSD), substance abuse, premature mortality, suicidal tendencies and various other ailme nts.[3,4,5,6,7,8,9,10] In addition to these serious consequences, veterans may suffer also prolonged feelings of loneliness.[11,12,13]

With the advent of the 21st century, the founder of Positive Psychology, Martin Seligman, contended that psychological investigations should focus more on acknowledging the positive aspects of human development than its negative aspects.[14,15] This positive perspective has been applied in numerous fields of psychology, such as substance abuse disorders in which the notion of recovery capital has emerged.[16,17] This concept of recovery brings into light the resources that one can muster when attempting to overcome a substance dependency. Similarly, the areas of stress management and trauma have seen an explosion in resiliency research[18,19] theories that focus on individuals' capacity to preserve and use their personal resources to counter the effect of adversity have gained attention as well,[20] such as the theories of hardiness,[21,22] physiological approach to resilience,[23] and conservation of resources.[24,25] Common to all these models is the investigation of both internal (e.g., physiological, personality traits) and external (e.g., social support, Socioeconomic Status (SES)) contributors to effective coping resources to reduce the deleterious effect of adversities. Anything can turn out to be an effective resource if it helps in achieving a goal,[26] but the challenge lies in identifying these 'anythings.' That is exactly what present day research and interventions on adversity focus on.

The concept of post–traumatic growth (PTG)[27] facilitated research in the domain of growth development after trauma. PTG has been defined as a positive psychological change that emerges after encountering traumatic or significantly difficult life events.[28] In the beginning, Janoff-Bulman worked on the emotional and cognitive processes of trauma survivors that facilitate the rebuilding of shattered assumptions about their world.[29] This early work paved the way for the concept of PTG. She came up with three types of PTG processes. In the first, one gains strength through suffering. In the second, there is an existential re-evaluation of one's life. Finally, the third involves psychological preparedness.[30] Psychological preparedness to rebuild one's assumptive world is similar to how communities rebuilds themselves post-earthquake.[31]

Building on Janoff–Bulman's work, Tedeschi and Calhoun developed a model of PTG, which has been revised over time on the basis of empirical findings. In 2004, Tedeschi and Calhoun carried out an empirical evaluation of the construct. They proposed five domains in which PTG can happen independently. Individuals experiencing PTG may not show growth in all five domains, but there should be significant gain in at least one of the following domains: (a) finding more meaning in relationships with others; (b) feeling an increase in personal strength; (c) identifying new possibilities in life; (d) increasing one's appreciation for life or drastically shifting priorities; and (e) growing in spiritual or existential matters.[32] These PTG factors have been found in different populations when empirically examined. Indeed, PTG was reported in cancer patients,[33] soldiers with military combat experience[34] and those who experienced the death of a loved one.[35] Meta-analyses of the PTG studies identified some PTG in 75 % to 90 % of the samples under study.[36,37] These findings suggest that PTG is a frequent and widespread consequence of trauma. There is even a possibility that PTG as an outcome is more prevalent than PTSD.

In Tedeschi and Calhoun's model, at the time of trauma, individuals are in an emotional state and must first manage their distress. At this point, the majority of people show symptoms similar to PTSD or acute stress disorder.

As time goes by, these symptoms subside for most people[38] as they engage in cognitive processing of the event and do healthy rumination which involves assimilation, accommodation and over accommodation of the traumatic event.[39] PTG is influenced by various contextual factors at the time of disclosure of the traumatic event[40]. Efforts to disclose, process and resolve the traumatic event are either facilitated or hindered by the reactions of others upon disclosure of the traumatic events and the socio cultural context in which the traumatic event occurred. Finally, personal disposition directly influences growth after trauma. For instance, resilience will impact how individuals recover post trauma.[41] Details regarding the process of PTG as given by Tedeschi and Calhoun are discussed next.

Process of PTG

Acute symptomology – Developing short-term symptoms after experiencing a traumatic event is a common occurrence. Symptoms commonly reported are automatic, intrusive and repetitive in nature, causing arousal, such that in the initial stage, individuals need to find a way to manage their symptoms.[42] PTG is dependent on some amount of personal distress experienced by the survivor but prolonged arousal and re-experiencing will exhaust individuals and, without intervention, they may end up developing a psychopathology.[43]

Rumination – Rumination can be described as repetitive thinking that may not always be intrusive. It involves reviewing, problem solving, attempting to find meaning in the traumatic event and, finally, looking for some positive outcomes out of this struggle.[44] This holistic description of the rumination process makes it appear as an adaptive process. However, trauma doesn't always result in positive outcomes. Remembering negative life experiences can overwhelm an individual and might trigger a fight or flight response. Clinicians prefer using a reflective rumination approach, which is more constructive and requires a conscious effort for the development of a new understanding of the world.[45]

Cognitive theorists have long recognized the process of rumination as assimilation, over accommodation/accommodation.[46] In the initial phase, an individual tries to assimilate the traumatic event into their cognitive framework. Attempts are made to alter one's perception to fit their prior beliefs. Besides assimilation, the individual indulges in accommodation when they alter their beliefs about the self and the world to gain control over their cognitions and feel safe. At times it leads to over accommodation where an individual generalizes his experience to an extent that may not be warranted. For instance, a person who has been assaulted by some they know may then find it exceedingly difficult to trust all other people. The goal of the therapy is to make an individual accommodate, gain cognitive flexibility to permit the transformation of existing beliefs and the addition of new knowledge.[47]

Social Sharing – Social sharing and support are important factors in promoting PTG and reducing unhealthy ruminative behavior.[48] Studies have shown that social support doesn't always contribute towards a positive change. What promotes PTG is not the number of individuals who can provide support but rather it is the survivor's perception of social support.[49] Social networks can helps individuals in reconstructing and reorganizing their broken belief system.[50] When a traumatic event is processed in a supportive environment, the association between the traumatic event and their negative response may be weakened. In addition, individuals may be exposed to multiple perspectives of their experience. Becoming aware of others' perspectives, especially those that are more positive than one's own, can give hope.

Social Reference – Socializing with individuals who have experienced a similar trauma can provide unique growth opportunities.[51] This is due to acquiring a better understanding of one's needs. Narrating a traumatic event to an empathetic audience can result in the validation of one's experience and the receipt of social support, which in turn help in growth and healing.[52] Social support is enhanced when it provided by group members who share similar attitudes, assumptions, and customs.[53] The bringing together of all survivors of a traumatic event is not only related to healthy cognitive processing, but also to the practice of healing activities, such as building memorials, indulging in altruistic behavior, and participating in public demonstrations targeting the prevalence of similar traumatic events.[54] Participating in such healing activities will, in turn, promote hope, trust, and solidarity and ultimately, well-being, empathy, positive affect, and increased interpersonal attraction.[55]

A military environment is definitely different from a civilian one. Indeed, the culture itself is dramatically different. Hence providing healing opportunities in terms of groups exclusively for military personnel may promote growth.

Human Narrative – An unprocessed traumatic event can be stored in our brain for decades, for example, Vietnam veterans still seek treatment for PTSD. The first step in the treatment of PTSD in military members is to provide the veteran with some vocabulary to describe the traumatic event and its related thoughts.[56] Human beings use language to create a meaning for daily life events, which, if required, can be rationalized, explained and challenged at a later time.[57] Self-narratives are shaped in the form of stories by organizing various cognitive and behavioral components. This process facilitates the development of one's identity as well as the perception of one's social environment.[58] Regardless of the reality of the objective world, our internal truth is determined by what we believe about ourselves, which, in turn influences our adjustment process. So an individual's narrative about a traumatic event is important in the treatment of PTSD as it can help them to repair a fragmented sense of identity and lead them towards the path of PTG.

Post – Traumatic Growth in Veterans

Tedeschi and Mc Nally[59] reviewed various PTG studies conducted on veterans. They summarized different ways to encourage veterans to self-disclose in a constructive manner. Such self-disclosure aids in the development of a coherent trauma narrative. The process of a trauma narrative can include various steps that can tap into any of the five domains of PTG, which relate to developing new goals and/or new ways of living[60]. Furthermore, PTG enables one to develop ways to deal with future challenges. Combat exposure may be a life changing experience for veterans. The quest for achieving a higher functioning level than before the trauma is a process rather than a goal.

Social support has emerged as one of the strongest factors contributing positively to the development of PTG in veterans. Many soldiers coming back from deployments report enhanced meaningful relationships with their unit members.[61] They believe that only those who have served in similar circumstances can understand their experiences. This belief constitutes a major obstacle in the reintegration of veterans into civilian life. Social support from fellow soldiers is most crucial in the development of PTG.[62] Actually, social support correlates significantly with both the extent of self-disclosure of and of the coherent narration of the trauma.[63] Not surprisingly, a crucial component of programs for veterans is the development of veterans' social support network and personal strengths[64].

Research outside the field of psychology suggests that adventure therapy helps in promoting pro-social communities that lead to increased social sharing and increased opportunities to practice newly acquired cognitions and behaviors.[65] Adventure therapy is defined as "the perspective use of adventure experiences provided by mental health professionals, often conducted in natural settings that kinesthetically engage clients on cognitive, affective and behavioral levels."[66] Results from a meta-analysis suggest that outdoor programs positively influence self-esteem, trust, team building, self-efficacy, affect, leadership skills, interpersonal development, and hope.[67] Many of these outcomes contribute towards PTG. Outdoor programs include seven important facets that undoubtedly contribute to the outcomes: being a motivated learner, taking on the role of instructor, participating in a physical and social outdoor environment, acquiring an adventure based experience, encountering success and finally transferring the learning to other environments.[68] Debriefings by instructors also help the participant to make sense of their experience[69]. For all these reasons, individuals previously exposed to combat trauma may benefit from such outdoor programs.

Kurl Hahn, the father of adventure training programs, developed an experiential school for preparing young men for sea battles during WWII. Hahn, an educator, believed that individuals need to learn from their weaknesses and failures to keep going when facing adversities. He thought it essential to develop an awareness of one's own subjective struggles. Through experiential activities, participants would not only discover their own selves but would discover unique ways to contribute to the greater need of the community.[70] Such adventure programs still exist; especially in the context of outdoor leadership programs.[71] Though not intended as a therapy, such programs nevertheless promote the development of personal strengths and provide life-enhancing opportunities.

The US Armed Forces started communicating with Outward Bound[72] in 1969 as they had common experiential training approach.[73] Participants of outward-bound veteran programs reported improved self-confidence, compassion, goal setting, self-actualization, group collaboration, effective communication, healthy and balanced life style, social responsibility, conflict resolution and problem solving immediately after attending the program.[74] Although such adventure programs were unsuccessful at decreasing PTSD symptoms, they nonetheless helped participants in rediscovering enjoyment in outdoors and other life activities, feeling more in control, overcoming negative emotions more easily, and in improving relationships with others.[75] Overall, the evidence suggests that various adventure based programs or outward-bound programs, if designed specifically for veterans or military personnel with combat exposure, may help in promoting PTG.

Resilience and Post- Traumatic Growth

Because resilience and PTG are both conducive to a positive adaptation post-trauma, they are often considered as being similar. Differences between the two constructs have been examined only rarely. Tedeschi and Calhoun were the first to distinguish between the two constructs. A resilient individual is able to thrive in extremely stressful situations. PTG and resilience differ on three aspects. Firstly, resilience involves an ability to preserve their pre-traumatic performance level, but to qualify as PTG, pre-traumatic performance level must be exceeded, at least in one of the five facets of PTG. Secondly, resilient individuals don't have to struggle much to recover from adversity whereas one must struggle before achieving PTG. Finally, resilient individuals don't look for a meaning in the traumatic event, whereas the process of PTG requires some cognitive processing.[76]

Resilience and PTG are definitely different constructs. Understanding the differences between these two constructs is important because it may influence what techniques counsellors choose to use with trauma survivors. Understanding the differences is especially critical when counselling soldiers who suffer from PTSD because PTSD is closely associated with both resilience and PTG and these relationships are complex. For example, one study found that prisoners of war and veterans who were high on resilience reported lower PTG than participants who reported high levels of PTSD symptoms.[77] The authors pointed out that those who reported higher levels of PTSD symptoms post-trauma also reported more positive psychological changes when they encountered further severe adversities.

Post – Traumatic Growth and Suicide

Although PTG may not act as a protective factor against suicide, it is inversely related to suicide.[78] Indeed, results of a study conducted with US army soldiers with combat experience identified an inverse relationship between suicide ideation and PTG.[79] Higher levels of PTG were associated with lower suicidal ideation. The Interpersonal Theory (IPT) model of suicide can help explain this inverse relationship through several mechanisms. As explained by the IPT model, one's desire to commit suicide is facilitated by two interpersonal constructs, thwarted belongingness and perceived burdensomeness. According to this theory the capability to indulge in suicidal behavior is different from the desire to commit suicide. PTG may protect from perceptions of burdensomeness, an important component of the IPT model. Such PTG factors as relating better to others, looking for new opportunities and increased personal strength may decrease the perception of carrying a burden. Relating better to others calls for increasing efforts in one's relationships, having faith that one can count on others during difficult times, accepting others' needs and feeling close to others. One no longer has to face burdens alone. Being able to identify new possibilities may aid in the development of faith that one can improve one's life circumstances, thereby reducing perceived burdensomeness. An increased awareness of one's personal strengths may lead to increased self-confidence that one is able to deal with adversities and that one is stronger than previously thought. These mechanisms may promote self-reliance and reduce perceived burdensomeness.

Another component of the IPT model is thwarted belongingness, which may also be countered by PTG factors such as relating better to others and spiritual change. PTG promotes a positive change in one's perception of relationships and provides as well as an increased sense of closeness with others. Such feelings develop a sense of belongingness, which, in turn, may help reduce the feeling of thwarted belongingness. Similarly, one's affiliation to a religion may increase one's sense of belongingness.[80] This is one way that spiritual change may counter thwarted belongingness. The PTG process involves the disclosure of the traumatic experience. A well-received disclosure may promote PTG[81] whereas a negative reaction to the disclosure may fuel the feeling of not being understood.

However, according to the IPT model, PTG cannot protect against the acquired capability of suicide. A change in pain tolerance and/or a reduction in the fear of death may influence suicide ideation. However, pain tolerance and fear of death seem unlikely to change, even after growth post trauma.

Although not directly influencing suicide intention, PTG, may indirectly help reduce suicidal ideation by promoting well-being.

Conclusion

PTG is an important construct that carries the implicit hope that adversities can help human beings rise like phoenixes. To promote post-traumatic growth in military personnel in the future, we recommend developing active programs similar to outward-bound programs. Programs developed to promote PTG in soldiers must take the unique nature of the military culture into consideration. Further, we need to recognize that PTG is not an end in itself, but rather an ongoing process. Growth-related activities need to be spread over a period of time. An individual's readiness for growth may vary in time and needs to be evaluated periodically because it can be deceptive, illusory, or real.

Notes

1 https://www.goodreads.com/author/quotes/2782.Viktor_E_Frankl
2 Jacob Stein Yale, Avigal Snir, Zahava Soloman, "When man harms man: The interpersonal ramifications of war captivity," In Katie E Cherry (Ed.), *Traumatic stress and long-term recovery*, pp. 113–132 (2015). New York: Springer.
3 Yuval Neria, Zahava Soloman, Karni Ginzberg, Rachel Dekel, Dan Enoch, and Abraham Ohry, "Post traumatic residues of captivity: A follow up of Israeli ex-prisoners of war," *Journal of Clinical Psychiatry*, 61, no. 1 (January 2000): 39–46, https://doi.org/10.4088/jcp.v61n0110

4 Thomas N. Dikel, Brien E. Engdahl, and Raina Eberly, "PTSD in former prisoners of war: Prewar, wartime, and post war factors," *Journal of Trauma Stress*, 18, no. 1 (February 2005): 69, https://doi.org/10.1002/jts.20002
5 Lante S. Rintamaki, Frances M. Weaver, Philip L. Elbaum, Edward N. Klama, and Scott A. Miskevics, "Persistance of traumatic memories in World War II prisoners of war," *Journal of the American Geriatrics Society*, 57, no. 12 (December 2009): 2257, https://doi.org/10.1111/j.1532-5415.2009.02608.x
6 Zahava Soloman, Talya Greene, Tsachi Eni-Dor, Gadi Zerach, Yael Benyamini, and Avi Ohry, "The long-term implications of war captivity for mortality and health," *Journal of Behavioural Medicine* 37, no. 5 (October 2014): 849, https://doi.org/10.1007/s10865-013-9544-3
7 Craig J. Bryan, James E. Griffith, Brian T. Pace, Kent Hinkson, AnnaBelle O. Bryan, Tracy A. Clemans, and Zac E. Imel, "Combat exposure and risk for suicidal thoughts and behaviours among military personnel and veterans: A systematic review and meta-analysis," *Suicide Life Threat Behaviour*, 45, no. 5 (October 2015), 655, https://doi.org/10.1111/sltb.12163
8 Jessica J. Fulton, Patrick S. Calhoun, H. Ryan Wagner, Amie R. Schry, Lauren P. Hair, Nicole Feeling, Eric Elbogen, and Jean C. Beckham, "The prevalence of posttraumatic stress disorder in Operation Enduring Freedom/Operation Iraqi Freedom (OEF/OIF) Veterans: A meta-analysis," *Journal of Anxiety Disorders*, 31 (April 2015): 98, https://doi.org/10.1016/j.janxdis.2015.02.003
9 Mark C. Russell, Charles R. Figley, Kriston R. Robertson, "Investigating the psychiatric lessons of war and pattern of preventable wartime behavioral health crises," *Journal of Psychology and Behavioural Science*, 3, (July 2015): 1, https://doi:10.15640/jpbs.v3n1a1
10 Chiao-Wen Lan, David A. Fiellin, Declan T. Barry, Kendall J. Bryant, Adam J. Gordon, E. Jennifer Edelman, Julie R. Gaither, Stephen A. Maisto, and Brandon D.L. Marshall, "The epidemiology of substance use disorders in US Veterans: A systematic review and analysis of assessment methods," *The American Journal on Addictions*, 25, no. 1(January 2016): 7, https://doi.org/10.1111/ajad.12319
11 Philipp Kuwert, Christine Knaevelsrud, and Robert H. Pietrzak, "Loneliness among older veterans in the United States: Results from the National Health and Resilience in Veterans Study," *Journal of the American Geriatrics Society*, 22, no. 6 (June 2014): 564, https://doi.org/10.1016/j.jagp.2013.02.013
12 Zahava Solomon, Moshe Bensimon, Talya Greene, Danny Horesh, Tsachi Ein-Dor, "Loneliness trajectories: The role of posttraumatic symptoms and social support," *Journal of Loss and Trauma*, 20, no. 1 (August 2014):1, https://doi/abs/10.1080/15325024.2013.815055
13 Jacob Y. Stein and Rivka Tuval-Mashiach, "Loneliness and isolation in life-stories of Israeli veterans of combat and captivity," *Psychological Trauma*, 7, no. 2 (March 2015): 122, https://doi.org/10.1037/a0036936.

14 Martin E.P. Seligman and Mihaly Csikszentmihalyi, "Positive psychology: An introduction," *American Psychologist*, 55, no. 1(February 2000): 5, https://doi:10.1037/0003-066X.55.1.5.

15 Martin E.P. Seligman, Tracy A Steen, Nansook Park, and Christopher Peterson, "Positive psychology progress: Empirical validation of interventions," *American Psychologist*, 60, no. 5 (August 2005): 410, https://doi.org/10.1037/0003-066X.60.5.410

16 Robert Granfield and William Cloud, *Coming clean: Overcoming addiction without treatment* (1999). New York: New York University Press,.

17 Emily A. Hennessy, "Recovery capital: A systematic review of the literature," *Addiction Research and Theory*, 25, no. 5 (March 2017): 349, https://doi.org/10.1080/16066359.2017.1297990

18 Anthony D. Mancini and George A. Bonnano, "Resilience in the face potential trauma," *Current Directions in Psychological Science*, 14, no. 3 (June 2005):135, https://doi:10.1111/j.0963-7214.2005.00347.x

19 George A. Bonnano, Maren Westphal, and Anthony Mancini, "Resilience to loss and potential trauma," *Annual Review of Clinical Psychology*, 7, no. 1 (April 2010): 511, https://doi.org/10.1146/annurev-clinpsy-032210-104526

20 Steven E. Hobfoll, "Social and Psychological resources and adaptation," *Review of General Psychology*, 6, no. 4 (December 2002): 307, https://doi.org/10.1037/1089-2680.6.4.307

21 S.C. Kobassa, "Stressful life events, personality and health: an inquiry into hardiness," *Journal of Personality and Social Psychology* 37, no. 1 (January 1979): 1, https://doi.org/10.1037//0022-3514.37.1.1

22 Kevin J. Eschleman, Nathan A. Bowling, and Gene M. Alarcon, "A meta-analytic examination of hardiness," *Internal Journal of Stress Management*, 17, no. 4 (2010): 277. https://doi.org/10.1037/a0020476.

23 Sarah R. Horn, Dennis Charney, and Adriana Feder, "Understanding resilience: New approaches for preventing and treating PTSD," *Experimental Neurology*, 284, No pt B (October 2016): 119, https://doi.org/10.1016/j.expneurol.2016.07.002

24 Steven E. Hobfoll, "Conservation of resources: A new attempt at conceptualising stress," *The American Psychologist*, 44, no. 3 (March 1989): 513, https://doi.org/10.1037//0003-066x.44.3.513

25 Steven E. Hobfoll, Vanessa Tirone, L. Holmgreen, and James I. Grehart, "Conservation of resources theory applied to major stress," In George Fink (Ed.), *Stress: Concepts, cognition, emotion and behavior*, p. 65 (2016). San Diego CA: Elsevier.

26 Jonnathan R.B. Halbesleben, Jean-Pierre Neveu, Samantha C. Paustian-Underdahl, and Mina Westman, "Getting to the "COR" understanding the role of resources in conservation of resources theory," *Journal of Management*, 40, no. 5 (July 2014): 1334, https://doi:10.1177/0149206314527130.

27 Richard G. Tedeschi and Lawrence G. Calhoun, "Posttraumatic growth: Conceptual foundations and empirical evidence," *Psychological Inquiry*, 4, no. 1 (2004):1, https://doi.org/10.1207/s15327965pli1501_01
28 Ibid.
29 Ronnie Janoff Bulman, *Shattered Assumptions: Towards a new psychology of trauma* (1992) New York: Free Press.
30 Ronnie Janoff Bulman, "Schema-change perspectives on posttraumatic growth," In Lawrence G. Calhoun, and Richard G. Tedeschi (Eds.), *The handbook of posttraumatic growth: Research and practice*, 81 (2006). Mahwah New Jersey: Lawrence Erlbaum, https://psycnet.apa.org/record/2006-05098-005
31 Lawrence G. Calhoun and Richard G. Tedeschi, "Beyond recovery from trauma: Implications for clinical practice and research," *Journal of Social Issus*, 54, no. 2 (April 1998): 357. https://doi.org/10.1111/0022-4537.701998070
32 Tedeschi and Calhoun, *"Posttraumatic growth"*
33 Miri Cohen and Maya Numa, "Posttraumatic growth in breast cancer survivors: a comparison of volunteers and non-volunteers," *Psycho-Oncology*, 20, no. 1 (January 2011): 69, https://doi.org/10.1002/pon.1709
34 Robert H. Pietraz, Marc B. Goldstein, James C. Malley, Allison J. Rivers, Douglas C. Johnson, Charles A. Morgan III, and Steven M. Southwick, "Posttraumatic growth in veterans of Operations Enduring Freedom and Iraqi Freedom," *Journal of Affective Disorders*, 126, no. 1–2 (October 2010): 230, https://doi.org/10.1016/j.jad.2010.03.021
35 Nick J Gerrish, Murray James Dynk, and Ali Marsh, "Post-traumatic growth and bereavement," *Mortality*, 14, no. 3 (August 2009): 226, https://doi:10.1080/13576270903017032
36 Richard G. Tedeschi, Crystal L. Park, and Lawrence G. Calhoun, *Posttraumatic growth: Positive changes in the aftermath of crisis* (1998). Mahwah New Jersey: Lawrence Erlbaum Associates.
37 John A. Updegraff and Shelley E. Taylor, "From vulnerability to growth: The positive and negative effects of stressful life events," In John Harvey and Eric Miller (Eds.), *Loss and trauma* (2000). Philadelphia, PA: Taylor & Francis.
38 Richard G. Tedeschi, "Posttraumatic growth in combat veterans," *Journal of Clinical Psychology in Medical Settings*, 18, no. 2 (June 2011): 137, https://pubmed.ncbi.nlm.nih.gov/21626349/
39 Lawrence G. Calhoun, Arnie Cann, and Richard G. Tedeschi, "The posttraumatic growth model: Socio-cultural considerations," In Tzipi Weiss and Roni Berger (Eds.), *Posttraumatic growth: A cross-cultural perspective*, p. 1 (2010). Hoboken New Jersey: John Wiley & Sons.
40 Ibid.
41 Ibid.
42 Tedeschi & Calhoun, *"Posttraumatic growth"*.

43 Ibid.
44 Lawrence G. Calhoun and Richard G. Tedeschi, "The foundations of posttraumatic growth: An expanded framework," In Lawrence G. Calhoun and Richard G. Tedeschi (Ed.), *Handbook of posttraumatic growth: Research and practice*, p. 1 (2006). Mahwah New Jersey: Lawrence Erlbaum Associates.
45 Susan Nolen-Hoeksema and Christopher G. Davis, "Theoretical and methodological issues in the assessment and interpretation of posttraumatic growth," *Psychological Inquiry*, 15, no. 1(2004): 60, https://www.jstor.org/stable/20447203
46 Kathleen M. Chard, Patricia A. Resick, Candice M. Monson, and Karen A. Kattar, *Cognitive processing therapy: Veteran/ military version* (2013). Washington DC: Department of Veterans Affairs, https://www.div12.org/wp-content/uploads/2014/11/Group-CPT-Manual.pdf
47 Ibid.
48 Tedeschi and Calhoun, *"Posttraumatic growth"*.
49 Manuel Berrera, "Distinctions between social support concepts, measures, and models," *American Journal of Psychology*, 14, no. 4 (January 1986): 413–445, https://doi:10.1007/BF00922627
50 R Tedeschi & Calhoun, *"Posttraumatic growth"*.
51 Shannon L. Currie, A. Day, and E. Kelloway, "Bringing the troops back home: Modeling the postdeployment reintegration experience," *Journal of Occupational Health Psychology*, 16, no. 1 (2011): 38–47, https://doi:10.1037/a0021724
52 Robert A. Neiymer, "Restoring loss: Fostering growth in the posttraumatic narrative," In Lawrence G. Calhoun and Richard G. Tedeschi (Eds.), *Handbook of posttraumatic growth: Research and practice*, pp. 68–80 (2006). Mahwah New Jersey: Lawrence Erlbaum Associates.
53 Lawrence G. Calhoun and Richard G. Tedeschi, "Posttraumatic growth in clinical perspective," in Lawrence G. Calhoun and Richard G. Tedeschi (Eds.), *Handbook of posttraumatic growth: Research and practice*, pp. 291–310, (2006). Mahwah New Jersey: Lawrence Erlbaum Associates.
54 Joseph de Rivera and Dareo Paez, "Emotional climate, human security, and cultures of peace," *Journal of Social Issues*, 63, no. 2 (March 2007): 233–253, https://www.ehu.eus/documents/1463215/1504269/de_Rivera_&_P%E2%80%A0ez_(2007).pdf
55 Dario Paez, Nekane Basabe, Silvia Ubillos, and Jose Luis Gonzalez-Castro, "Social sharing, participation in demonstrations, emotional climate, and coping with collective violence after the March 11th Madrid Bombings1," *Journal of Social Issues*, 63, no. 2 (June 2007): 323–337, https://doi:10.1111/j.1540-4560.2007.00511.x
56 Robert A. Neiymer, "Restoring loss: Fostering growth in the posttraumatic narrative," In Lawrence G. Calhoun and Richard G. Tedeschi (eds.), *Handbook*

of posttraumatic growth: Research and Practice, (Mahwah New Jersey: Lawrence Erlbaum Associates, 2006), 68–80.
57 Donald Meichenbaum, "Resilience and Posttraumatic growth," in *Handbook of Posttraumatic growth: Research and Practice*, ed, Lawrence G Calhoun and Richard G Tedeschi (Mahwah New Jersey: Lawrence Erlbaum Associates, 2006), 355–368, https://psycnet.apa.org/record/2006-05098-018
58 Robert A Neiymer, "Restoring loss: Fostering growth in the posttraumatic narrative," In Lawrence G. Calhoun and Richard G. Tedeschi (Eds.), *Handbook of posttraumatic growth: Research and Practice*, pp. 68–80 (2006). Mahwah New Jersey: Lawrence Erlbaum Associates.
59 Richard G. Tedeschi and Richard J. McNally, "Can we facilitate growth in combat veterans?," *American Psychologist*, 66, no. 1 (January 2011): 19–24, https://psycnet.apa.org/doiLanding?doi=10.1037 %2Fa0021896
60 Richard G. Tedeschi and Lawrence G. Calhoun, "Beyond the concept of recovery: Growth and the experience of loss," *Death Studies*, 32, no. 1 (January 2008): 27–39, https://doi10.1080/07481180701741251
61 Donald Meichenbaum, *Roadmap to resilience: A guide for military, trauma victims and their families* (UK: Crown House Publishing, 2012).
62 Shira Maguen, Dawne S. Vogt, Lynda A. King, Daniel W. King, and Brett T. Litz, "Posttraumatic growth among Gulf War I Veterans: The predictive role of deployment-related experiences and background characteristics," *Journal of Loss and Trauma*, 11 (February 2007): 377–388, https://www.tandfonline.com/doi/10.1080/15325020600672004
63 Richard G Tedeschi, "Posttraumatic growth in combat veterans," *Journal of Clinical Psychology in Medical Settings*, 18, no. 2 (June 2011): 137, https://pubmed.ncbi.nlm.nih.gov/21626349/
64 Nansook Park, "Strengths and challenges during peace and war," *American Psychologist*, 66, no. 1 (January 2011): 65–72, https://pubmed.ncbi.nlm.nih.gov/21219050/
65 Michael A. Gass, H. L. Lee Gillis, and Keith C. Russell, *Adventure therapy: Theory, research and practice* (2012). New York: Routledge.
66 Ibid, p. 1.
67 Lee H. Gillis and Elizabeth Speelman, "Are challenge (ropes) courses an effective tool? a meta-analysis," *Journal of Experiential Education*, 31, no. 2 (2008): 111–135, http://citeseerx.ist.psu.edu/viewdoc/download?doi=10.1.1.564.4336&rep=rep1&type=pdf
68 Michael A. Gass, Lee H. Gillis, and Keith C. Russell, *Adventure therapy: Theory, research and practice* (2012). New York: Routledge.
69 Angela M. Passarelli, Eric E. Hall, and Mallory Anderson, "A strength-based approach to outdoor and adventure education: Possibilities for personal growth," *Journal of Experiential Education*, 33, no. 2 (September 2010): 120–135, http://citeseerx.ist.psu.edu/viewdoc/download?doi=10.1.1.564.4336&rep=rep1&type=pdf.

70 Joshua L. Miner and Joseph R. Boldt, *Outward bound USA: Crew not passengers* (2002). Seattle, WA: Mountaineers.
71 Outward Bound USA. (2021). Retrieved from: www.outwardbound.org
72 Outward Bound was established in 1962 as an American provider of outdoor education programs designed to encourage the achievement of personal potential in young people.
73 Outward Bound Veteran Program. (2021). Retrieved from: www.outwardbound.org
74 Alan Ewert, Marieke Van Puymbroeck, Jon Frankel, and Jillisa Overholt, "Adventure education and returning military veteran: What do we know?," *The Journal of Experiential Education*, 33, no. 4. (May 2011): 365–369, https://journals.sagepub.com/doi/10.1177/105382591003300408
75 Lee Hyer, Stephanie Boyd, Ray Scurfield, Dale Smith, and Jim Burke, "Effects of Outward Bound experience as an adjunct to inpatient PTSD treatment of war veterans," *Journal of Clinical Psychology*, 52, no. 3 (May 1996): 263–278, https://pubmed.ncbi.nlm.nih.gov/8835688/
76 Maren Westphal and George A Bonnano, "Posttraumatic growth and resilience to trauma: Different sides of the same coin or different coins?" *Applied Psychology: An International Review*, 56, no. 3 (July 2007): 417–427, https://iaap-journals.onlinelibrary.wiley.com/doi/10.1111/j.1464-0597.2007.00298.x
77 Gadi Zerach, Zahava Solomon, Assaf Cohen, and Tsachi Ein-Dor, "PTSD, resilence and post traumatic growth among ex-prisoners of war and combat veterans," *The Israel Journal of Psychiatry and Related Science*, 50, no. 2 (2013): 91–99, https://pubmed.ncbi.nlm.nih.gov/24225436/
78 Nigel E. Bush, Nancy A. Skopp, Russell McCann, and David D. Luxton, "Posttraumatic growth against suicidal ideation after deployment and combat exposure," *Military Medicine*, 176, no. 11 (November 2011): 1215–1222, https://pubmed.ncbi.nlm.nih.gov/22165648/
79 Shayne M. Gallaway, Amy M. Millikan, and Michael R. Bell, "The association between deployment related posttraumatic growth among U S army soldiers and negative behavioural health conditions," *Journal of Clinical Psychology*, 67, no. 12(December 2011): 1151–1160, https://pubmed.ncbi.nlm.nih.gov/22042556/
80 Chaeyoon Lim and Robert D. Putnam "Religion, social networks, and life satisfaction," *American Sociological Review*, 75, no. 6 (December 2010): 913–933, https://www.jstor.org/stable/25782172
81 Cassie M. Lindstorm, Arnie Cann, Lawrence G. Calhoun, and Richard G. Tedeschi, "The relationship of core belief challenge, rumination, disclosure, and sociocultural elements to posttraumatic growth," *Psychological Trauma*, 5, no. 1 (2013): 150–155, https://psycnet.apa.org/record/2011-06103-001

The Authors

Hubert Annen, Ph.D., is the head of Military Psychology and Pedagogy Studies at Military Academy/ETH Zurich and the head of the Swiss Army assessment centres for prospective Defence Attachés, General Staff Officers, and Professional Officers and NCOs respectively. Parallel to his professional career, he is also still active as a reserve officer in the Swiss Armed Forces with the rank of colonel. His research interests include the evaluation and validation of assessment and selection procedures for military leaders, motivational aspects in the military context, military education, military values and virtues, and the trainability and measurability of individual resilience.

Sophie Arana, Ph.D., is a senior scientist within the UK Ministry of Defence Army Health and Performance Research team, primarily tasked with conducting and delivering research to improve the health and performance of servicewomen through-career. She currently leads on projects in nutrition and postpartum return to occupational fitness, and is a member of the Servicewomen's Health Improvement Focus Team; Defence's women's health initiative to improve policy and education for servicewomen. She is a member of the Defence Nutrition Advisory Service, is a UK representative on several NATO panels and holds an honorary senior research fellow position with the University College London.

Petrus C. Bester, D Phil, Leadership in Performance and Change, is a senior lecturer in Industrial Psychology (Mil) at the South African Military Academy, Faculty of Military Science at Stellenbosch University. He joined academia after a career in Defence Intelligence and the South African Military Health Service. His research interests include the application of psychology within the national security context, military leadership, military ethics, integrity, performance enhancement and test construction.

Danielle Charbonneau, Ph.D., is an emeritus professor at the Royal Military College where she was a member of the department of Military Psychology and Leadership for 25 years. Her research interest included mainly emotional intelligence, transformational leadership, resilience and military culture.

Lobna Chérif, PhD, CAPP, CRT, PPCC, MBSP, is an Associate Professor in the Department of Military Psychology and Leadership and the Chair in Resilience at the Royal Military College of Canada. She has a PhD in Psychology, is a

Certified Applied Positive Psychology Practitioner, a Certified Resilience Trainer, a Professional and Personal Certified Coach, and a certified Mindfulness-Based Strengths Practitioner. Her current research focuses on character strengths, resilience, and accomplishment – and how to apply this understanding to improving well-being and performance.

Abhijit Prakash Deshpande, Ph.D., is the Director, Board of University Development and a Professor of Management at Symbiosis International (Deemed University) in Pune, India. He is involved in academic audits at the University and is a member of the Board of Studies and a Ph.D. Research Guide under the Faculty of Management; formerly an ex-member of the Board of Management, and Lecturer in the Department of Business Studies at Bahrain Training Institute (Kingdom of Bahrain). His major areas of scholarly interest and expertise are in human resource management and strategic management. His doctoral thesis he completed in Human Resource Management on the topic of Application of the Competency Modeling Approach with Special Reference to Select Private Sector Organizations. He has authored a book, contributed to book chapters, and has publications in Scopus and Web of science-indexed journals.

Tyler Freeman, Ph.D., is a Cognitive Psychologist and Director of Army Research at ICF Inc. At ICF, Dr. Freeman provides applied research and consulting services to United States military organizations. Prior to joining ICF, he was a Consortium Research Fellow at the U.S. Army Research Institute. Dr. Freeman employs a research-to-practice approach to support the development of Army leaders and teams. His research interests include assessment development and validation; development and testing of applied training products; leadership, and team performance. Dr. Freeman has authored papers and delivered conference presentations on a variety of military-relevant topics such as stress mindsets, military learning organizations, leader influence, team cognition, and critical thinking.

Walter Giusti holds graduate and postgraduate qualifications in psychology from Monash University, Australia. He has worked as a senior analyst and assistant director for a number of Australian government agencies, including the Department of Defence Inspector-General Division where he undertook high level Defence-wide performance reviews, and the Mental Health, Psychology and Rehabilitation Branch of Joint Health Command conducting research on the Defence Force PULSE – a personnel climate assessment tool. He continued work in this area for Australian government justice agencies, particularly through the development of the Justice Working Well Profile. Walter's research interests include the application of behavioural and psychometric profiling in the military, national security and justice sectors.

Martin I. Jones, Ph.D., is a sport and exercise psychologist with over 20 years of experience working with elite performers in sports, business, and the military. Martin has an MSc and PhD in sport psychology from Loughborough University, is an honorary professor in performance psychology at Hartpury University, and is currently studying for an MSc in Sleep Medicine at Oxford University. Martin spent more than ten years working in higher education in the UK and Canada, he received several awards for teaching and research excellence and published more than 40 peer-reviewed papers on human performance topics. Martin is the Principal Advisor on the DSTL Optimising Human Performance research project, and the UK's lead member on the NATO Human Factors and Medicine (HFM) panel.

Vrishti Kapoor, M.Sc. in Clinical Psychology, is a practicing psychologist and founder at Psyog Wellness, a visiting psychologist at Dr. Rohit Garg's Mind Vriksha, Delhi, and a consultant psychologist at Dr. Era Dutta's Mind Wellness, Kolkata, India. Following her keen interest in holistic well-being of self and others, she recently completed Basic Teacher's Training course (200 hours) in Yoga Education at The Yoga Institute, Delhi, India. Her research interests include holistic well-being, positive psychology, childhood trauma, yoga and mental health, yoga as a therapeutic tool, and therapeutic effects of yoga.

Edith Knight, LCol, C.D., M.Sc., fills the role of Personnel Capabilities manager within the Canadian Army Headquarters. Since joining the Regular force in 2003, she has worked in various capacities on a wide array of projects, including: training development, program evaluation, recruiting and selection system design, job analyses, scale measurement and validation, survey design and strategy development. Her research interests have included fatigue, well-being, meaningful work and flight safety climate and culture.

Updesh Kumar, Ph.D., is Scientist 'G' (R) and former Head, Mental Health Division at Defense Institute of Psychological Research (DIPR), R & D Organization (DRDO), Ministry of Defense, Delhi, India. Presently, he is associated with Rashtriya Raksha University (RRU), an institution of national importance, Pioneer National Security, and the Police University of India, Gandhinagar, Gujarat. After obtaining his doctorate in suicidal behavior from

Punjab University, Chandigarh, India, he has more than 30 years of experience as a Scientist in an R&D organization. He specializes in personality assessment, suicidal behavior, health psychology, and military psychology. As a military psychologist, Dr. Kumar has been involved in selecting commissioned rank officers at various Services Selection Boards (SSBs) and was responsible for monitoring the entire selection system of the Indian Armed Forces for around three decades. In addition, as chief instructor, he has imparted training in psychological techniques to service officers and scientists manning the various SSBs and conducted many counselling techniques training programs and M.Sc. military psychology courses for the officers of the Indian Armed Forces and Central Armed Police Forces (CAPF). As internationally acclaimed author and editor, Dr. Kumar has edited among other books "The Routledge International Handbook of Military Psychology and Mental Health (Routledge, 2020, selected for the 2022 edition of Doody's Core Titles®) and most recently "Emotion, Well-Being, and Resilience: Theoretical Perspectives and Practical Applications (CRC Press & Apple Academic Press, 2021). He has been the past editor of the Journal of Indian Academy of Applied Psychology (JIAAP) and a reviewer with many international and national journals of repute. Dr. Updesh Kumar is the recipient of many awards including *DRDO Best Popular Science Communication Award* by Government of India, 2009 and *DRDO's Scientist of the Year Award* (2013) by Government of India.

Jürgen Léon, M.Sc., is a clinical psychologist at the Klinik Teufen Group in Switzerland. He has a master's degree in clinical psychology, and has conducted a scientific project on Leadership at the Military Academy at ETH Zürich. He is currently working on post-traumatic growth and Burn-out resilience by leaders. Specifically, he investigates possibilities and challenges of interventions with attention to when adversity transpires across leaders' careers.

Allister MacIntyre, Ph.D., is a Psychology Professor, and former Department Head at the Royal Military College of Canada (RMC). In 2015 he was elected to RMC's Senate and served in this capacity until 2021. He spent 31 years in uniform as a military psychologist with the Canadian Forces and has participated on several international leadership and psychology panels. The co-editor of several books, he has been an executive member of the Canadian Psychological Association since 2002. His long history with the International Military Testing Association (IMTA) commenced in 1995 and in 2011 he was the recipient of IMTA's Harry Greer Award in recognition of his contributions to the organization. In 2015 he was elected as a Director with IMTA's Management Board and, in 2021, he was elected as the Chair of IMTA's Management Board. His scholarly activities focus primarily on leadership and ethics and, from 2017 until 2022, he co-chaired a multi-nation research project for NATO examining ethical leadership within NATO and Partners for Peace (PfP) countries.

Madlaina Niederhauser, M.Sc., is project staff member at the Department for Military Psychology and Pedagogy at the Military Academy at ETH Zurich. Her research interests include individual resilience and its trainability.

Samir Rawat, Lt Col, Ph.D., is a combat veteran and a cognitive-behavioural psychologist from India who has extensive experience in strengthening intrapersonal and interpersonal skills development with individuals and among teams; his focused work areas include executive, life and performance coaching, mentoring, behavioural counselling, soft skills as well as outbound training, and OD consulting. He is a decorated soldier with proven leadership skills in Siachen Glacier, which, at 21000 ft, is the highest battlefield in the world where he commanded an Infantry rifle company as a volunteer Officer. He was also conferred a President's gallantry award for recapturing enemy held positions during Kargil operations in 1999. With a Ph.D. in Psychology, Masters in Management and a MPhil in Defence & Strategic Studies, Samir brings with him over three decades of experience in training and human resource optimisation. In 2021, he was nominated as the first Indian to be an eminent state of art keynote speaker at the 32nd International Congress of Psychology 2021(ICP) in Prague which is the biggest international event in the psychological world with more than 9000 psychologists conglomerating for an event that happens once every four years. Besides providing learning and development initiatives in diverse domains across verticals, Samir Rawat is concurrently pursuing his doctoral research in management with an emphasis on self-regulation, teamwork and leadership.

Nity Sharma, PhD, is Scientist 'E' at Defence Institute of Psychological Research (DIPR), Defence Research and Development Organization (DRDO), Ministry of Defence (MoD), Government of India, at Delhi. Besides doing research work, she also imparts training to service officers of Indian Armed Forces and scientists of DRDO in the area of personality assessment for personnel selection, stress management and mindfulness techniques. Her research interests include psychological well-being through mindfulness techniques, emotion management and regulation, personality assessment, military psychology, positive psychology and psychological test development. She is the recipient of Scientist of the Year Award 2022, conferred by DRDO, for her outstanding contribution in scientific research.

Gayle Sherwell is currently working in a senior occupational health and safety role for a major Australian government agency. She has extensive experience in designing and implementing workforce health and safety strategies and management systems, including employee mental well-being initiatives, for complex and diverse organizations. Gayle's more recent work in this area has been for the government justice sector.

Verma Swati, Ph.D. is Scientist 'D' from Defence Research and Development organization, Ministry of Defence, Government of India. Currently she is working as senior psychologist at Naval Selection Board, Visakhapatnam. She has been awarded commendation by commanding in chief eastern naval command in December 2016. She has been actively involved in the selection of naval officers for Indian Navy. She has published several papers in different journals of international repute and presented her research work at various national and international conferences. Recently she wrote a chapter on 'Transition from Military to Civilian life' in Routledge's 'Handbook of Military Psychology and Mental Health'.

Liisi Toom, MA, is a practicing performance coach (ICF) and mental performance enhancement skills consultant for elite athletes, executives and teams working in high-risk, high-performance contexts. She worked as a lecturer of psychology in Estonian Military Academy (2020–2022), where she focused on developing interventions to enhance resilience and performance as well as introduced coaching as an effective tool for military leaders. Her work and research interests include crisis communication, building strategies, psychological performance enhancement of individual high-performers and teams as well as integration of resilience to company culture and individual life.

Elena Trentini, M.Sc., is a Ph.D. student at the University of Lausanne (Switzerland). She has a master's degree in clinical psychology, and she is currently working on personality and emotion regulation strategies. Specifically, she investigates the efficiency of emotion regulation strategies in relationship with personality at the experiential, expressive, and physiological levels.

Dakshi Walia has completed an M.Sc. in Health Psychology from Centre for Health Psychology, School of Medical Sciences, University of Hyderabad, India.

Before that she completed B.A. (Hons) in Psychology from University of Delhi, India. She has published in social sciences journals and presented papers at national and international conferences. She is the recipient of Professor Deepak Bhat best paper presentation award in the 54th National & 23rd International Conference of Indian Academy of Applied Psychology (IAAP) held at Department of Psychology, Kurukshetra University, Kurukshetra, India.

Valerie Wood, Ph.D., is a Research and Evaluation Specialist for the Road to Mental Readiness Program, within the Canadian Forces Health Services. Her previous research has focused on the role of adult attachment in relationship conflict and spousal adjustment to military deployments, relationships and public health, and the well-being and resilience of military recruits and members.

Studies in Military Psychology and Pedagogy

Editor: Hubert Annen

Band 1 Edmund A. van Trotsenburg: Militärpädagogik. 1989.

Band 2 Hermann Jung / Heinz Florian: Grundlagen der Militärpädagogik. Eine Anleitung zu pädagogisch verantwortetem Handeln. 1994.

Band 3 Franz Kernic: Demokratie und Wehrsystem. Aufsätze zum Verhältnis von Gesellschaft, politischem System und Heer in Österreich. 1997.

Band 4 Edwin R. Micewski: Grenzen der Gewalt – Grenzen der Gewaltlosigkeit. Zur Begründung der Gewaltproblematik im Kontext philosophischer Ethik und politischer Philosophie. 1998.

Band 5 Rudolf Egger / Heinz Florian (Hrsg.): Pädagogische Professionalisierung im Bundesheer. Dokumentation und Reflexion des PädAk-Sonderstudienganges Wehrpädagogisches Management. 1999.

Band 6 Franz Kernic / Harald Haas: Warriors for Peace. A Sociological Study on the Austrian Experience of UN Peacekeeping. 1999.

Band 7 Franz Kernic / Jean M. Callaghan / Philippe Manigart: Public Opinion on European Security and Defense. A Survey of European Trends and Public Attitudes Toward CFSP and ESDP. 2002.

Band 8 Heinz Florian (ed.): Military Pedagogy – An International Survey. 2002.

Band 9 Edwin R. Micewski / Hubert Annen (eds.): Military Ethics in Professional Military Education – Revisited. 2005.

Band 10 Hubert Annen / Wolfgang Royl (eds.): Military Pedagogy in Progress. 2007.

Band 11 Hubert Annen / Wolfgang Royl (eds.): Educational Challenges Regarding Military Action. 2010.

Band 12 Hubert Annen / Juha Mäkinen / Can Nakkas (eds.): Thinking and Acting in Military Pedagogy. 2013.

Band 13 Duraid Jalili / Hubert Annen (eds.): Professional Military Education. A Cross-Cultural Survey. 2019.

Band 14 Nadine Eggimann Zanetti: Values and Virtues in the Military. 2020.

Band 15 Allister MacIntyre, Danielle Charbonneau and Hubert Annen (eds.): Positive Psychology in the Military. 2023.

www.peterlang.de